GACE Elementary Education Math (501) Worksheets

2024

A Comprehensive Review of GACE Math Test

By

Reza Nazari

All inquiries should be addressed to:
info@effortlessMath.com
www.EffortlessMath.com

Published by: **Effortless Math Education Inc.**

For Online Math Practice Visit www.EffortlessMath.com

Welcome to
GACE Math
2024

Thank you for choosing Effortless Math for your GACE Elementary Education Math test preparation and congratulations on making the decision to take the GACE Math test! It's a remarkable move you are taking, one that shouldn't be diminished in any capacity. That's why you need to use every tool possible to ensure you succeed on the test with the highest possible score, and this extensive study guide is one such tool.

If math has never been a strong subject for you, don't worry! This book will help you prepare for (and even ACE) the GACE Elementary Education Math. As test day draws nearer, effective preparation becomes increasingly more important. Thankfully, you have this comprehensive study guide to help you get ready for the test. With this guide, you can feel confident that you will be more than ready for the GACE Elementary Education Math test when the time comes.

First and foremost, it is important to note that this book is a study guide and not a textbook. It is best read from cover to cover. Every lesson of this "self-guided math book" was carefully developed to ensure that you are making the most effective use of your time while preparing for the test. This up-to-date guide reflects the 2024 test guidelines and will put you on the right track to hone your math skills, overcome exam anxiety, and boost your confidence, so that you can have your best to succeed on the GACE Math test.

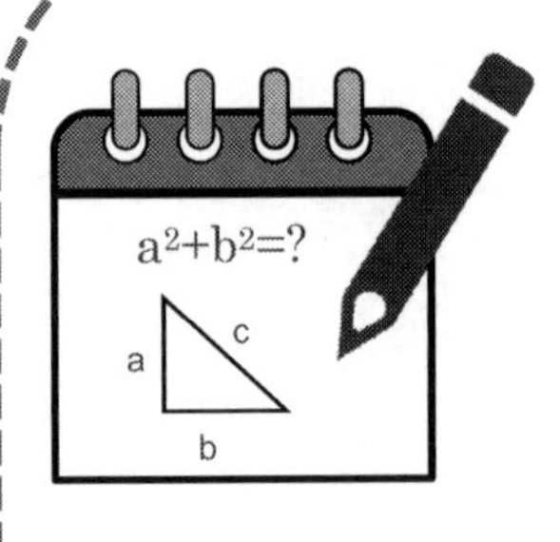

This study guide will:

☑ Explain the format of the GACE Elementary Education Math test.

☑ Describe specific test-taking strategies that you can use on the test.

☑ Provide GACE Math test-taking tips.

☑ Review all GACE Math concepts and topics you will be tested on.

☑ Help you identify the areas in which you need to concentrate your study time.

☑ Offer exercises that help you develop the basic math skills you will learn in each section.

☑ Give **2 realistic and full-length practice tests** (featuring new question types) with detailed answers to help you measure your exam readiness and build confidence.

This resource contains everything you will ever need to succeed on the GACE Math test. You'll get in-depth instructions on every math topic as well as tips and techniques on how to answer each question type. You'll also get plenty of practice questions to boost your test-taking confidence. In addition, in the following pages you'll find:

- **How to Use This Book Effectively** – This section provides you with step-by-step instructions on how to get the most out of this comprehensive study guide.

- **How to study for the GACE Math Test** – A six-step study program has been developed to help you make the best use of this book and prepare for your GACE Math test. Here you'll find tips and strategies to guide your study program and help you understand GACE Math and how to ace the test.

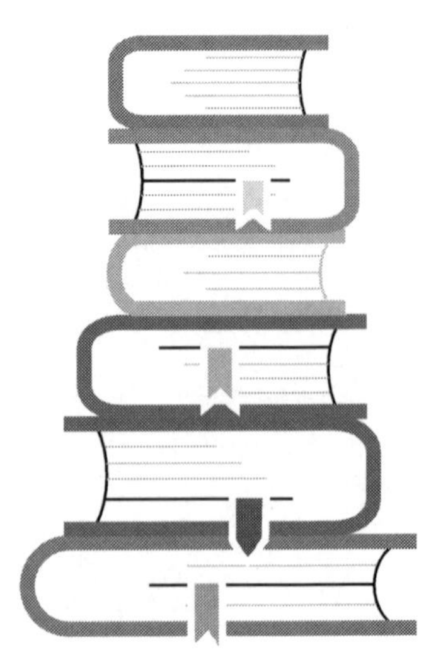

- **GACE Elementary Education Math Review** – Learn everything you need to know about the GACE Math test.
- **GACE Elementary Education Math Test-Taking Strategies** – Learn how to effectively put these recommended test-taking techniques into use for improving your GACE Math score.
- **Test Day Tips** – Review these tips to make sure you will do your best when the big day comes.

Effortless Math's GACE Math Online Center

Effortless Math Online GACE Math Center offers a complete study program, including the following:

- ✓ Step-by-step instructions on how to prepare for the GACE Math test
- ✓ Numerous GACE Math worksheets to help you measure your math skills
- ✓ Complete list of GACE Math formulas
- ✓ Video lessons for all GACE Math topics
- ✓ Full-length GACE Math practice tests
- ✓ And much more…

No Registration Required.

Visit **EffortlessMath.com/GACE** to find your online GACE Math resources.

How to Use This Book Effectively

Look no further when you need a study guide to improve your math skills to succeed on the math portion of the GACE Math test. Each chapter of this comprehensive guide to the GACE Math will provide you with the knowledge, tools, and understanding needed for every topic covered on the test.

It's imperative that you understand each topic before moving onto another one, as that's the way to guarantee your success. Each chapter provides you with examples and a step-by-step guide of every concept to better understand the content that will be on the test. To get the best possible results from this book:

- **Begin studying long before your test date**. This provides you ample time to learn the different math concepts. The earlier you begin studying for the test, the sharper your skills will be. Do not procrastinate! Provide yourself with plenty of time to learn the concepts and feel comfortable that you understand them when your test date arrives.
- **Practice consistently**. Study GACE Math concepts at least 20 to 30 minutes a day. Remember, slow and steady wins the race, which can be applied to preparing for the GACE Math test. Instead of cramming to tackle everything at once, be patient and learn the math topics in short bursts.
- Whenever you get a math problem wrong, **mark it off, and review it later** to make sure you understand the concept.
- Start each session by **looking over the previous material.**
- Once you've reviewed the book's lessons, **take a practice test at the back of the book** to gauge your level of readiness. Then, review your results. Read detailed answers and solutions for each question you missed.
- **Take another practice test** to get an idea of how ready you are to take the actual exam. Taking the practice tests will give you the confidence you need on test day. Simulate the GACE Math testing environment by sitting in a
- quiet room free from distraction. Make sure to clock yourself with a timer.

How to Study for the GACE Math Test

Studying for the GACE Math test can be a really daunting and boring task. What's the best way to go about it? Is there a certain study method that works better than others? Well, studying for the GACE Math can be done effectively. The following six-step program has been designed to make preparing for the GACE Math test more efficient and less overwhelming.

Step 1 - Create a study plan
Step 2 - Choose your study resources
Step 3 - Review, Learn, Practice
Step 4 - Learn and practice test-taking strategies
Step 5 - Learn the GACE Math Test format and take practice tests
Step 6 - Analyze your performance

STEP 1: Create a Study Plan

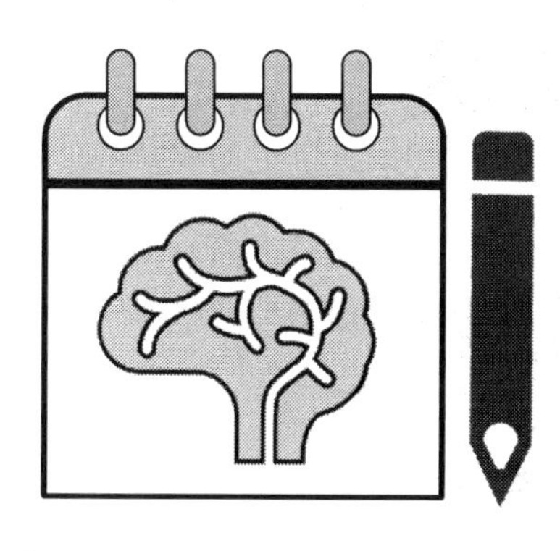

It's always easier to get things done when you have a plan. Creating a study plan for the GACE Math test can help you to stay on track with your studies. It's important to sit down and prepare a study plan with what works with your life, work, and any other obligations you may have. Devote enough time each day to studying. It's also a great idea to break down each section of the exam into blocks and study one concept at a time.

It's important to understand that there is no "right" way to create a study plan. Your study plan will be personalized based on your specific needs and learning style.

Follow these guidelines to create an effective study plan for your GACE Math test:

★ **Analyze your learning style and study habits** – Everyone has a different learning style. It is essential to embrace your individuality and the unique way you learn. Think about what works and what doesn't work for you. Do you prefer GACE Math prep books or a combination of textbooks

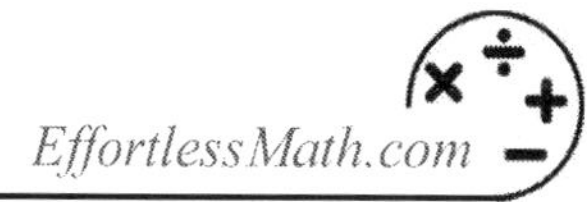

and video lessons? Does it work better for you if you study every night for thirty minutes or is it more effective to study in the morning before going to work?

★ **Evaluate your schedule** – Review your current schedule and find out how much time you can consistently devote to GACE Math study.

★ **Develop a schedule** – Now it's time to add your study schedule to your calendar like any other obligation. Schedule time for study, practice, and review. Plan out which topic you will study on which day to ensure that you're devoting enough time to each concept. Develop a study plan that is mindful, realistic, and flexible.

★ **Stick to your schedule** – A study plan is only effective when it is followed consistently. You should try to develop a study plan that you can follow for the length of your study program.

★ **Evaluate your study plan and adjust as needed** – Sometimes you need to adjust your plan when you have new commitments. Check in with yourself regularly to make sure that you're not falling behind in your study plan. Remember, the most important thing is sticking to your plan. Your study plan is all about helping you be more productive. If you find that your study plan is not as effective as you want, don't get discouraged. It's okay to make changes as you figure out what works best for you.

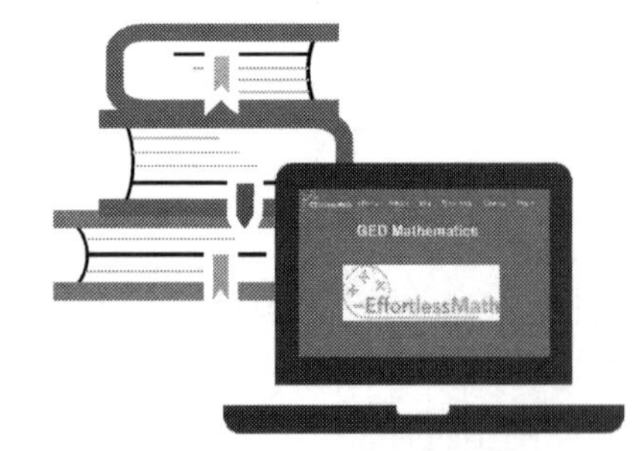

STEP 2: Choose Your Study Resources

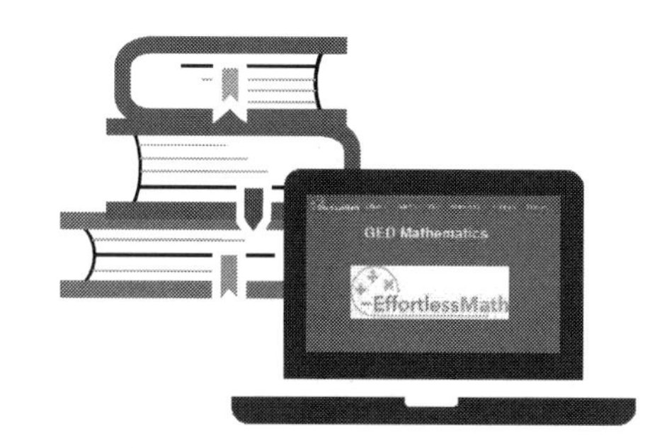

There are numerous textbooks and online resources available for the GACE Math test, and it may not be clear where to begin. Don't worry! This study guide provides everything you need to fully prepare for your GACE Math test. In addition to the book content, you can also use Effortless Math's online resources. (video lessons, worksheets, formulas, etc.) On each page, there is a link (and a QR code) to an online webpage which provides a comprehensive review of the topic, step-by-step instruction, video tutorial, and numerous examples and exercises to help you fully understand the concept.

Simply visit EffortlessMath.com/GACE to find your online GACE Math resources.

STEP 3: Review, Learn, Practice

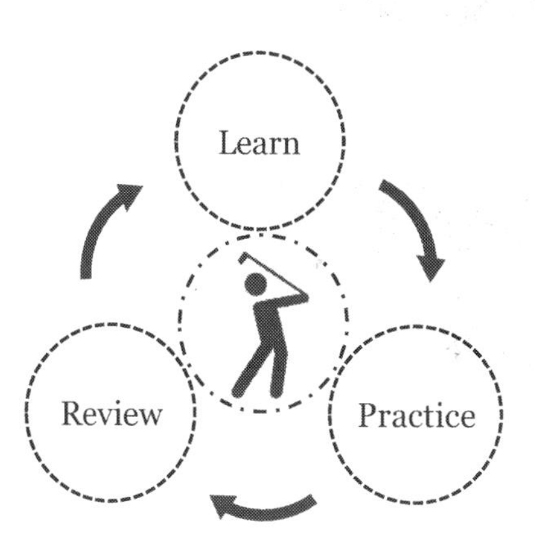

This GACE Math study guide breaks down each subject into specific skills or content areas. For instance, the percent concept is separated into different topics–percent calculation, percent increase and decrease, percent problems, etc. Use this book to help you go over all key math concepts and topics on the GACE Math test. As you read each chapter, take notes or highlight the concepts you would like to go over again in the future. If you're unfamiliar with a topic or something is difficult for you, do additional research on it. For each math topic, plenty of instructions, step-by-step guides, and examples are provided to ensure you get a good grasp of the material. You can also find video lessons on the Effortless Math website for each GACE Math concept.

Quickly review the topics you do understand to get a brush-up of the material. Be sure to do the practice questions provided at the end of every chapter to measure your understanding of the concepts.

Step 4: Learn and Practice Test-taking Strategies

In the following sections, you will find important test-taking strategies and tips that can help you earn extra points. You'll learn how to think strategically and when to guess if you don't know the answer to a question. Using GACE Math test-taking strategies and tips can help you raise your score and do well on the test. Apply test taking strategies on the practice tests to help you boost your confidence.

Step 5: Learn the GACE Elementary Education Math Test Format and Take Practice Tests

The GACE Math Test Review section provides information about the structure of the GACE Math test. Read this section to learn more about the GACE Math test structure, different test sections, the number of questions in each section, and the section time limits. When you have a prior understanding of the test format and different types of GACE Math questions, you'll feel more confident when you take the actual exam.

Once you have read through the instructions and lessons and feel like you are ready to go – take advantage of both of the full-length GACE Math practice tests available in this study guide. Use the practice tests to sharpen your skills and build confidence.

The GACE Math practice tests offered at the end of the book are formatted similarly to the actual GACE Math test. When you take each practice test, try to simulate actual testing conditions. To take the practice tests, sit in a quiet space, time yourself, and work through as many of the questions as time allows. The practice tests are followed by detailed answer explanations to help you find your weak areas, learn from your mistakes, and raise your GACE Math score.

Step 6: Analyze Your Performance

After taking the practice tests, look over the answer keys and explanations to learn which questions you answered correctly and which you did not. Never be discouraged if you make a few mistakes. See them as a learning opportunity. This will highlight your strengths and weaknesses.

You can use the results to determine if you need additional practice or if you are ready to take the actual GACE Math test.

Looking for more?

Visit EffortlessMath.com/GACE to find hundreds of GACE Math worksheets, video tutorials, practice tests, GACE Math formulas, and much more.

Or scan this QR code. 

No Registration Required.

GACE Test Review

The GACE Math test is an essential benchmark for individuals aiming to embark on a career in teaching at the elementary level within the state of Georgia. This examination is designed with a comprehensive approach to assess the foundational knowledge and teaching skills required to meet the high standards of elementary education. The test ensures that candidates are well-equipped to create an engaging and effective learning environment for young learners.

The GACE Math test is divided into two subtests, covering a wide array of subjects essential for providing a holistic elementary education:

- Reading and Language Arts
- Social Studies
- Analysis
- Mathematics
- Science
- Health Education, Physical Education, and the Arts

Focusing on the Mathematics section of the GACE Math test, the areas of knowledge include:

- counting and cardinality
- operations and algebraic thinking
- numbers and operations in base 10
- numbers and fractions
- measurement concepts and data

This framework aims to prepare prospective elementary educators with a solid theoretical and practical background in mathematics. It is designed to enable educators to cultivate an environment that promotes mathematical curiosity and competence among elementary students, laying down a strong foundation for their future academic and personal growth.

Contents

Simplifying Fractions

Simplify each fraction.

1) $\frac{10}{15} =$

2) $\frac{8}{20} =$

3) $\frac{12}{42} =$

4) $\frac{5}{20} =$

5) $\frac{6}{18} =$

6) $\frac{18}{27} =$

7) $\frac{15}{55} =$

8) $\frac{24}{54} =$

9) $\frac{63}{72} =$

10) $\frac{40}{64} =$

11) $\frac{23}{46} =$

12) $\frac{35}{63} =$

13) $\frac{32}{36} =$

14) $\frac{81}{99} =$

15) $\frac{16}{64} =$

16) $\frac{14}{35} =$

17) $\frac{19}{38} =$

18) $\frac{18}{54} =$

19) $\frac{56}{70} =$

20) $\frac{40}{45} =$

21) $\frac{9}{90} =$

22) $\frac{20}{25} =$

23) $\frac{32}{48} =$

24) $\frac{7}{49} =$

25) $\frac{18}{48} =$

26) $\frac{54}{108} =$

Simplifying Fractions - Answers

Simplify each fraction.

1) $\frac{10}{15} = \frac{2}{3}$

2) $\frac{8}{20} = \frac{2}{5}$

3) $\frac{12}{42} = \frac{2}{7}$

4) $\frac{5}{20} = \frac{1}{4}$

5) $\frac{6}{18} = \frac{1}{3}$

6) $\frac{18}{27} = \frac{2}{3}$

7) $\frac{15}{55} = \frac{3}{11}$

8) $\frac{24}{54} = \frac{4}{9}$

9) $\frac{63}{72} = \frac{7}{8}$

10) $\frac{40}{64} = \frac{5}{8}$

11) $\frac{23}{46} = \frac{1}{2}$

12) $\frac{35}{63} = \frac{5}{9}$

13) $\frac{32}{36} = \frac{8}{9}$

14) $\frac{81}{99} = \frac{9}{11}$

15) $\frac{16}{64} = \frac{1}{4}$

16) $\frac{14}{35} = \frac{2}{5}$

17) $\frac{19}{38} = \frac{1}{2}$

18) $\frac{18}{54} = \frac{1}{3}$

19) $\frac{56}{70} = \frac{4}{5}$

20) $\frac{40}{45} = \frac{8}{9}$

21) $\frac{9}{90} = \frac{1}{10}$

22) $\frac{20}{25} = \frac{4}{5}$

23) $\frac{32}{48} = \frac{2}{3}$

24) $\frac{7}{49} = \frac{1}{7}$

25) $\frac{18}{48} = \frac{3}{8}$

26) $\frac{54}{108} = \frac{1}{2}$

Adding and Subtracting Fractions

Calculate and write the answer in the lowest term.

1) $\frac{1}{5}+\frac{1}{7}=$

2) $\frac{3}{7}+\frac{4}{5}=$

3) $\frac{3}{8}-\frac{1}{9}=$

4) $\frac{4}{5}-\frac{5}{9}=$

5) $\frac{2}{9}+\frac{1}{3}=$

6) $\frac{3}{10}+\frac{2}{5}=$

7) $\frac{9}{10}-\frac{4}{5}=$

8) $\frac{7}{9}-\frac{3}{7}=$

9) $\frac{3}{4}+\frac{1}{3}=$

10) $\frac{3}{8}+\frac{2}{5}=$

11) $\frac{3}{4}-\frac{2}{5}=$

12) $\frac{7}{9}-\frac{2}{3}=$

13) $\frac{4}{9}+\frac{5}{6}=$

14) $\frac{2}{3}+\frac{1}{4}=$

15) $\frac{9}{10}-\frac{3}{5}=$

16) $\frac{7}{12}-\frac{1}{2}=$

17) $\frac{4}{5}+\frac{2}{3}=$

18) $\frac{5}{7}+\frac{1}{5}=$

19) $\frac{5}{9}-\frac{2}{5}=$

20) $\frac{3}{5}-\frac{2}{9}=$

21) $\frac{7}{9}+\frac{1}{7}=$

22) $\frac{5}{8}+\frac{2}{3}=$

23) $\frac{4}{7}+\frac{2}{3}=$

24) $\frac{6}{7}-\frac{4}{9}=$

25) $\frac{4}{5}-\frac{2}{15}=$

26) $\frac{2}{9}+\frac{4}{5}=$

Adding and Subtracting Fractions - Answers

 ***Calculate and write the answer in the lowest term.***

1) $\frac{1}{5}+\frac{1}{7}=\frac{12}{35}$

2) $\frac{3}{7}+\frac{4}{5}=\frac{43}{35}$

3) $\frac{3}{8}-\frac{1}{9}=\frac{19}{72}$

4) $\frac{4}{5}-\frac{5}{9}=\frac{11}{45}$

5) $\frac{2}{9}+\frac{1}{3}=\frac{5}{9}$

6) $\frac{3}{10}+\frac{2}{5}=\frac{7}{10}$

7) $\frac{9}{10}-\frac{4}{5}=\frac{1}{10}$

8) $\frac{7}{9}-\frac{3}{7}=\frac{22}{63}$

9) $\frac{3}{4}+\frac{1}{3}=\frac{13}{12}$

10) $\frac{3}{8}+\frac{2}{5}=\frac{31}{40}$

11) $\frac{3}{4}-\frac{2}{5}=\frac{7}{20}$

12) $\frac{7}{9}-\frac{2}{3}=\frac{1}{9}$

13) $\frac{4}{9}+\frac{5}{6}=\frac{23}{18}$

14) $\frac{2}{3}+\frac{1}{4}=\frac{11}{12}$

15) $\frac{9}{10}-\frac{3}{5}=\frac{3}{10}$

16) $\frac{7}{12}-\frac{1}{2}=\frac{1}{12}$

17) $\frac{4}{5}+\frac{2}{3}=\frac{22}{15}$

18) $\frac{5}{7}+\frac{1}{5}=\frac{32}{35}$

19) $\frac{5}{9}-\frac{2}{5}=\frac{7}{45}$

20) $\frac{3}{5}-\frac{2}{9}=\frac{17}{45}$

21) $\frac{7}{9}+\frac{1}{7}=\frac{58}{63}$

22) $\frac{5}{8}+\frac{2}{3}=\frac{31}{24}$

23) $\frac{4}{7}+\frac{2}{3}=\frac{26}{21}$

24) $\frac{6}{7}-\frac{4}{9}=\frac{26}{63}$

25) $\frac{4}{5}-\frac{2}{15}=\frac{2}{3}$

26) $\frac{2}{9}+\frac{4}{5}=\frac{46}{45}$

Multiplying and dividing fraction

Solve and write the answer in lowest term.

1) $\frac{1}{2} \times \frac{4}{5} =$

2) $\frac{1}{5} \times \frac{6}{7} =$

3) $\frac{1}{3} \div \frac{1}{7} =$

4) $\frac{1}{7} \div \frac{3}{8} =$

5) $\frac{2}{3} \times \frac{4}{7} =$

6) $\frac{5}{7} \times \frac{3}{4} =$

7) $\frac{2}{5} \div \frac{3}{7} =$

8) $\frac{3}{7} \div \frac{5}{8} =$

9) $\frac{3}{8} \times \frac{4}{7} =$

10) $\frac{2}{9} \times \frac{6}{11} =$

11) $\frac{1}{10} \div \frac{3}{8} =$

12) $\frac{3}{10} \div \frac{4}{5} =$

13) $\frac{6}{7} \times \frac{4}{9} =$

14) $\frac{3}{7} \times \frac{5}{6} =$

15) $\frac{7}{9} \div \frac{6}{11} =$

16) $\frac{1}{15} \div \frac{2}{3} =$

17) $\frac{1}{13} \times \frac{1}{2} =$

18) $\frac{1}{12} \times \frac{4}{7} =$

19) $\frac{1}{15} \div \frac{4}{9} =$

20) $\frac{1}{16} \div \frac{1}{2} =$

21) $\frac{4}{7} \times \frac{5}{8} =$

22) $\frac{1}{11} \times \frac{4}{5} =$

23) $\frac{1}{18} \div \frac{5}{6} =$

24) $\frac{1}{15} \div \frac{3}{8} =$

25) $\frac{1}{11} \times \frac{3}{4} =$

26) $\frac{1}{14} \times \frac{2}{3} =$

Multiplying and Dividing Fractions - Answers

 ***Solve and write the answer in lowest terms.***

1) $\frac{1}{2} \times \frac{4}{5} = \frac{2}{5}$

2) $\frac{1}{5} \times \frac{6}{7} = \frac{6}{35}$

3) $\frac{1}{3} \div \frac{1}{7} = \frac{7}{3}$

4) $\frac{1}{7} \div \frac{3}{8} = \frac{8}{21}$

5) $\frac{2}{3} \times \frac{4}{7} = \frac{8}{21}$

6) $\frac{5}{7} \times \frac{3}{4} = \frac{15}{28}$

7) $\frac{2}{5} \div \frac{3}{7} = \frac{14}{15}$

8) $\frac{3}{7} \div \frac{5}{8} = \frac{24}{35}$

9) $\frac{3}{8} \times \frac{4}{7} = \frac{3}{14}$

10) $\frac{2}{9} \times \frac{6}{11} = \frac{4}{33}$

11) $\frac{1}{10} \div \frac{3}{8} = \frac{4}{15}$

12) $\frac{3}{10} \div \frac{4}{5} = \frac{3}{8}$

13) $\frac{6}{7} \times \frac{4}{9} = \frac{8}{21}$

14) $\frac{3}{7} \times \frac{5}{6} = \frac{5}{14}$

15) $\frac{7}{9} \div \frac{6}{11} = \frac{77}{54}$

16) $\frac{1}{15} \div \frac{2}{3} = \frac{1}{10}$

17) $\frac{1}{13} \times \frac{1}{2} = \frac{1}{26}$

18) $\frac{1}{12} \times \frac{4}{7} = \frac{1}{21}$

19) $\frac{1}{15} \div \frac{4}{9} = \frac{3}{20}$

20) $\frac{1}{16} \div \frac{1}{2} = \frac{1}{8}$

21) $\frac{4}{7} \times \frac{5}{8} = \frac{5}{14}$

22) $\frac{1}{11} \times \frac{4}{5} = \frac{4}{55}$

23) $\frac{1}{18} \div \frac{5}{6} = \frac{1}{15}$

24) $\frac{1}{15} \div \frac{3}{8} = \frac{8}{45}$

25) $\frac{1}{11} \times \frac{3}{4} = \frac{3}{44}$

26) $\frac{1}{14} \times \frac{2}{3} = \frac{1}{21}$

Adding Mixed Numbers

Solve and write the answer in lowest terms.

1) $3\frac{1}{5}+2\frac{2}{9}=$

2) $1\frac{1}{7}+5\frac{2}{5}=$

3) $4\frac{4}{5}+1\frac{2}{7}=$

4) $2\frac{4}{7}+2\frac{3}{5}=$

5) $1\frac{5}{6}+1\frac{2}{5}=$

6) $3\frac{5}{7}+1\frac{2}{9}=$

7) $3\frac{5}{8}+2\frac{1}{3}=$

8) $1\frac{6}{7}+3\frac{2}{9}=$

9) $2\frac{5}{9}+1\frac{1}{4}=$

10) $3\frac{7}{9}+2\frac{5}{6}=$

11) $2\frac{1}{10}+2\frac{2}{5}=$

12) $1\frac{3}{10}+3\frac{4}{5}=$

13) $3\frac{1}{12}+2\frac{1}{3}=$

14) $5\frac{1}{11}+1\frac{1}{2}=$

15) $3\frac{1}{21}+2\frac{2}{3}=$

16) $4\frac{1}{24}+1\frac{5}{8}=$

17) $2\frac{1}{25}+3\frac{3}{5}=$

18) $3\frac{1}{15}+2\frac{2}{10}=$

19) $5\frac{6}{7}+2\frac{1}{3}=$

20) $2\frac{1}{8}+3\frac{3}{4}=$

21) $2\frac{5}{7}+2\frac{2}{21}=$

22) $4\frac{1}{6}+1\frac{4}{5}=$

23) $3\frac{5}{6}+1\frac{2}{7}=$

24) $2\frac{7}{8}+3\frac{1}{3}=$

25) $3\frac{1}{17}+1\frac{1}{2}=$

26) $1\frac{1}{18}+1\frac{4}{9}=$

Adding Mixed Numbers - Answers

Solve and write the answer in lowest terms.

1) $3\frac{1}{5}+2\frac{2}{9}=5\frac{19}{45}$

2) $1\frac{1}{7}+5\frac{2}{5}=6\frac{19}{35}$

3) $4\frac{4}{5}+1\frac{2}{7}=6\frac{3}{35}$

4) $2\frac{4}{7}+2\frac{3}{5}=5\frac{6}{35}$

5) $1\frac{5}{6}+1\frac{2}{5}=3\frac{7}{30}$

6) $3\frac{5}{7}+1\frac{2}{9}=4\frac{59}{63}$

7) $3\frac{5}{8}+2\frac{1}{3}=5\frac{23}{24}$

8) $1\frac{6}{7}+3\frac{2}{9}=5\frac{5}{63}$

9) $2\frac{5}{9}+1\frac{1}{4}=3\frac{29}{36}$

10) $3\frac{7}{9}+2\frac{5}{6}=6\frac{11}{18}$

11) $2\frac{1}{10}+2\frac{2}{5}=4\frac{1}{2}$

12) $1\frac{3}{10}+3\frac{4}{5}=5\frac{1}{10}$

13) $3\frac{1}{12}+2\frac{1}{3}=5\frac{5}{12}$

14) $5\frac{1}{11}+1\frac{1}{2}=6\frac{13}{22}$

15) $3\frac{1}{21}+2\frac{2}{3}=5\frac{5}{7}$

16) $4\frac{1}{24}+1\frac{5}{8}=5\frac{2}{3}$

17) $2\frac{1}{25}+3\frac{3}{5}=5\frac{16}{25}$

18) $3\frac{1}{15}+2\frac{2}{10}=5\frac{4}{15}$

19) $5\frac{6}{7}+2\frac{1}{3}=8\frac{4}{21}$

20) $2\frac{1}{8}+3\frac{3}{4}=5\frac{7}{8}$

21) $2\frac{5}{7}+2\frac{2}{21}=4\frac{17}{21}$

22) $4\frac{1}{6}+1\frac{4}{5}=5\frac{29}{30}$

23) $3\frac{5}{6}+1\frac{2}{7}=5\frac{5}{42}$

24) $2\frac{7}{8}+3\frac{1}{3}=6\frac{5}{24}$

25) $3\frac{1}{17}+1\frac{1}{2}=4\frac{19}{34}$

26) $1\frac{1}{18}+1\frac{4}{9}=2\frac{1}{2}$

Subtracting Mixed Numbers

Solve and write the answer in lowest terms.

1) $3\frac{2}{5} - 1\frac{2}{9} =$

2) $5\frac{3}{5} - 1\frac{1}{7} =$

3) $4\frac{2}{5} - 2\frac{2}{7} =$

4) $8\frac{3}{4} - 2\frac{1}{8} =$

5) $9\frac{5}{7} - 7\frac{4}{21} =$

6) $11\frac{7}{12} - 9\frac{5}{6} =$

7) $9\frac{5}{9} - 8\frac{1}{8} =$

8) $13\frac{7}{9} - 11\frac{3}{7} =$

9) $8\frac{7}{12} - 7\frac{3}{8} =$

10) $11\frac{5}{9} - 9\frac{1}{4} =$

11) $6\frac{5}{6} - 2\frac{2}{9} =$

12) $5\frac{7}{8} - 4\frac{1}{3} =$

13) $9\frac{5}{8} - 8\frac{1}{2} =$

14) $4\frac{9}{16} - 2\frac{1}{4} =$

15) $3\frac{2}{3} - 1\frac{2}{15} =$

16) $5\frac{1}{2} - 4\frac{2}{17} =$

17) $5\frac{6}{7} - 2\frac{1}{3} =$

18) $3\frac{3}{7} - 2\frac{2}{21} =$

19) $7\frac{3}{10} - 5\frac{2}{15} =$

20) $4\frac{5}{6} - 2\frac{2}{9} =$

21) $6\frac{3}{7} - 2\frac{2}{9} =$

22) $7\frac{4}{5} - 6\frac{3}{7} =$

23) $10\frac{2}{3} - 9\frac{5}{8} =$

24) $9\frac{3}{4} - 7\frac{4}{9} =$

25) $15\frac{4}{5} - 13\frac{12}{25} =$

26) $13\frac{5}{12} - 7\frac{5}{24} =$

Subtracting Mixed Numbers - Answers

Solve and write the answer in lowest terms.

1) $3\frac{2}{5} - 1\frac{2}{9} = 2\frac{8}{45}$

2) $5\frac{3}{5} - 1\frac{1}{7} = 4\frac{16}{35}$

3) $4\frac{2}{5} - 2\frac{2}{7} = 2\frac{4}{35}$

4) $8\frac{3}{4} - 2\frac{1}{8} = 6\frac{5}{8}$

5) $9\frac{5}{7} - 7\frac{4}{21} = 2\frac{11}{21}$

6) $11\frac{7}{12} - 9\frac{5}{6} = 1\frac{3}{4}$

7) $9\frac{5}{9} - 8\frac{1}{8} = 1\frac{31}{72}$

8) $13\frac{7}{9} - 11\frac{3}{7} = 2\frac{22}{63}$

9) $8\frac{7}{12} - 7\frac{3}{8} = 1\frac{5}{24}$

10) $11\frac{5}{9} - 9\frac{1}{4} = 2\frac{11}{36}$

11) $6\frac{5}{6} - 2\frac{2}{9} = 4\frac{11}{18}$

12) $5\frac{7}{8} - 4\frac{1}{3} = 1\frac{13}{24}$

13) $9\frac{5}{8} - 8\frac{1}{2} = 1\frac{1}{8}$

14) $4\frac{9}{16} - 2\frac{1}{4} = 2\frac{5}{16}$

15) $3\frac{2}{3} - 1\frac{2}{15} = 2\frac{8}{15}$

16) $5\frac{1}{2} - 4\frac{2}{17} = 1\frac{13}{34}$

17) $5\frac{6}{7} - 2\frac{1}{3} = 3\frac{11}{21}$

18) $3\frac{3}{7} - 2\frac{2}{21} = 1\frac{1}{3}$

19) $7\frac{3}{10} - 5\frac{2}{15} = 2\frac{1}{6}$

20) $4\frac{5}{6} - 2\frac{2}{9} = 2\frac{11}{18}$

21) $6\frac{3}{7} - 2\frac{2}{9} = 4\frac{13}{63}$

22) $7\frac{4}{5} - 6\frac{3}{7} = 1\frac{13}{35}$

23) $10\frac{2}{3} - 9\frac{5}{8} = 1\frac{1}{24}$

24) $9\frac{3}{4} - 7\frac{4}{9} = 2\frac{11}{36}$

25) $15\frac{4}{5} - 13\frac{12}{25} = 2\frac{8}{25}$

26) $13\frac{5}{12} - 7\frac{5}{24} = 6\frac{5}{24}$

Multiplying Mixed Numbers

Solve and write the answer in lowest terms.

1) $1\frac{1}{8} \times 1\frac{3}{4} =$

2) $3\frac{1}{5} \times 2\frac{2}{7} =$

3) $2\frac{1}{8} \times 1\frac{2}{9} =$

4) $2\frac{3}{8} \times 2\frac{2}{5} =$

5) $1\frac{1}{2} \times 5\frac{2}{3} =$

6) $3\frac{1}{2} \times 6\frac{2}{3} =$

7) $9\frac{1}{2} \times 2\frac{1}{6} =$

8) $2\frac{5}{8} \times 8\frac{3}{5} =$

9) $3\frac{4}{5} \times 4\frac{2}{3} =$

10) $5\frac{1}{3} \times 2\frac{2}{7} =$

11) $6\frac{1}{3} \times 3\frac{3}{4} =$

12) $7\frac{2}{3} \times 1\frac{8}{9} =$

13) $8\frac{1}{2} \times 2\frac{1}{6} =$

14) $4\frac{1}{5} \times 8\frac{2}{3} =$

15) $3\frac{1}{8} \times 5\frac{2}{3} =$

16) $2\frac{2}{7} \times 6\frac{2}{5} =$

17) $2\frac{3}{8} \times 7\frac{2}{3} =$

18) $1\frac{7}{8} \times 8\frac{2}{3} =$

19) $9\frac{1}{2} \times 3\frac{1}{5} =$

20) $2\frac{5}{8} \times 4\frac{1}{3} =$

21) $6\frac{1}{3} \times 3\frac{2}{5} =$

22) $5\frac{3}{4} \times 2\frac{2}{7} =$

23) $9\frac{1}{4} \times 2\frac{1}{3} =$

24) $3\frac{3}{7} \times 7\frac{2}{5} =$

25) $4\frac{1}{4} \times 3\frac{2}{5} =$

26) $7\frac{2}{3} \times 3\frac{2}{5} =$

Multiplying Mixed Numbers – Answers

Solve and write the answer in lowest terms.

1) $1\frac{1}{8} \times 1\frac{3}{4} = 1\frac{31}{32}$

2) $3\frac{1}{5} \times 2\frac{2}{7} = 7\frac{11}{35}$

3) $2\frac{1}{8} \times 1\frac{2}{9} = 2\frac{43}{72}$

4) $2\frac{3}{8} \times 2\frac{2}{5} = 5\frac{7}{10}$

5) $1\frac{1}{2} \times 5\frac{2}{3} = 8\frac{1}{2}$

6) $3\frac{1}{2} \times 6\frac{2}{3} = 23\frac{1}{3}$

7) $9\frac{1}{2} \times 2\frac{1}{6} = 20\frac{7}{12}$

8) $2\frac{5}{8} \times 8\frac{3}{5} = 22\frac{23}{40}$

9) $3\frac{4}{5} \times 4\frac{2}{3} = 17\frac{11}{15}$

10) $5\frac{1}{3} \times 2\frac{2}{7} = 12\frac{4}{21}$

11) $6\frac{1}{3} \times 3\frac{3}{4} = 23\frac{3}{4}$

12) $7\frac{2}{3} \times 1\frac{8}{9} = 14\frac{13}{27}$

13) $8\frac{1}{2} \times 2\frac{1}{6} = 18\frac{5}{12}$

14) $4\frac{1}{5} \times 8\frac{2}{3} = 36\frac{2}{5}$

15) $3\frac{1}{8} \times 5\frac{2}{3} = 17\frac{17}{24}$

16) $2\frac{2}{7} \times 6\frac{2}{5} = 14\frac{22}{35}$

17) $2\frac{3}{8} \times 7\frac{2}{3} = 18\frac{5}{24}$

18) $1\frac{7}{8} \times 8\frac{2}{3} = 16\frac{1}{4}$

19) $9\frac{1}{2} \times 3\frac{1}{5} = 30\frac{2}{5}$

20) $2\frac{5}{8} \times 4\frac{1}{3} = 11\frac{3}{8}$

21) $6\frac{1}{3} \times 3\frac{2}{5} = 21\frac{8}{15}$

22) $5\frac{3}{4} \times 2\frac{2}{7} = 13\frac{1}{7}$

23) $9\frac{1}{4} \times 2\frac{1}{3} = 21\frac{7}{12}$

24) $3\frac{3}{7} \times 7\frac{2}{5} = 25\frac{13}{35}$

25) $4\frac{1}{4} \times 3\frac{2}{5} = 14\frac{9}{20}$

26) $7\frac{2}{3} \times 3\frac{2}{5} = 26\frac{1}{15}$

Dividing Mixed Numbers

Solve and write the answer in lowest terms.

1) $9\frac{1}{2} \div 2\frac{3}{5} =$

2) $2\frac{3}{8} \div 1\frac{2}{5} =$

3) $5\frac{3}{4} \div 2\frac{2}{7} =$

4) $8\frac{1}{3} \div 4\frac{1}{4} =$

5) $7\frac{2}{5} \div 3\frac{3}{4} =$

6) $2\frac{4}{5} \div 3\frac{2}{3} =$

7) $8\frac{3}{5} \div 4\frac{3}{4} =$

8) $6\frac{3}{4} \div 2\frac{2}{9} =$

9) $5\frac{2}{7} \div 2\frac{2}{9} =$

10) $2\frac{2}{5} \div 3\frac{3}{5} =$

11) $4\frac{3}{7} \div 1\frac{7}{8} =$

12) $2\frac{5}{7} \div 2\frac{4}{5} =$

13) $8\frac{3}{5} \div 6\frac{1}{5} =$

14) $2\frac{5}{8} \div 1\frac{8}{9} =$

15) $5\frac{6}{7} \div 2\frac{3}{4} =$

16) $1\frac{3}{5} \div 2\frac{3}{8} =$

17) $5\frac{3}{4} \div 3\frac{2}{5} =$

18) $2\frac{3}{4} \div 3\frac{1}{5} =$

19) $3\frac{2}{3} \div 1\frac{2}{5} =$

20) $4\frac{1}{4} \div 2\frac{2}{3} =$

21) $3\frac{5}{6} \div 2\frac{4}{5} =$

22) $2\frac{1}{8} \div 1\frac{3}{4} =$

23) $5\frac{1}{2} \div 2\frac{2}{5} =$

24) $3\frac{4}{7} \div 2\frac{2}{3} =$

25) $2\frac{4}{5} \div 3\frac{5}{6} =$

26) $2\frac{3}{7} \div 3\frac{2}{3} =$

Dividing Mixed Number – Answers

Solve and write the answer in lowest terms

1) $9\frac{1}{2} \div 2\frac{3}{5} = 3\frac{17}{26}$

2) $2\frac{3}{8} \div 1\frac{2}{5} = 1\frac{39}{56}$

3) $5\frac{3}{4} \div 2\frac{2}{7} = 2\frac{33}{64}$

4) $8\frac{1}{3} \div 4\frac{1}{4} = 1\frac{49}{51}$

5) $7\frac{2}{5} \div 3\frac{3}{4} = 1\frac{73}{75}$

6) $2\frac{4}{5} \div 3\frac{2}{3} = \frac{42}{55}$

7) $8\frac{3}{5} \div 4\frac{3}{4} = 1\frac{77}{95}$

8) $6\frac{3}{4} \div 2\frac{2}{9} = 3\frac{3}{80}$

9) $5\frac{2}{7} \div 2\frac{2}{9} = 2\frac{53}{140}$

10) $2\frac{2}{5} \div 3\frac{3}{5} = \frac{2}{3}$

11) $4\frac{3}{7} \div 1\frac{7}{8} = 2\frac{38}{105}$

12) $2\frac{5}{7} \div 2\frac{4}{5} = \frac{95}{98}$

13) $8\frac{3}{5} \div 6\frac{1}{5} = 1\frac{12}{31}$

14) $2\frac{5}{8} \div 1\frac{8}{9} = 1\frac{53}{136}$

15) $5\frac{6}{7} \div 2\frac{3}{4} = 2\frac{10}{77}$

16) $1\frac{3}{5} \div 2\frac{3}{8} = \frac{64}{95}$

17) $5\frac{3}{4} \div 3\frac{2}{5} = 1\frac{47}{68}$

18) $2\frac{3}{4} \div 3\frac{1}{5} = \frac{55}{64}$

19) $3\frac{2}{3} \div 1\frac{2}{5} = 2\frac{13}{21}$

20) $4\frac{1}{4} \div 2\frac{2}{3} = 1\frac{19}{32}$

21) $3\frac{5}{6} \div 2\frac{4}{5} = 1\frac{31}{84}$

22) $2\frac{1}{8} \div 1\frac{3}{4} = 1\frac{3}{14}$

23) $5\frac{1}{2} \div 2\frac{2}{5} = 2\frac{7}{24}$

24) $3\frac{4}{7} \div 2\frac{2}{3} = 1\frac{19}{56}$

25) $2\frac{4}{5} \div 3\frac{5}{6} = \frac{84}{115}$

26) $2\frac{3}{7} \div 3\frac{2}{3} = \frac{51}{77}$

Comparing Decimals

Compare. Use >, =, and <

1) 0.88 □ 0.088
2) 0.56 □ 0.57
3) 0.99 □ 0.89
4) 1.55 □ 1.65
5) 1.58 □ 1.75
6) 2.91 □ 2.85
7) 14.56 □ 1.456
8) 17.85 □ 17.89
9) 21.52 □ 21.052
10) 11.12 □11.03
11) 9.650 □ 9.65
12) 8.578 □ 8.568
13) 3.15 □ 0.315
14) 16.61 □ 16.16
15) 18.581 □ 8.991
16) 25.05 □ 2.505
17) 4.55 □ 4.65
18) 0.158 □ 1.58
19) 0.881 □ 0.871
20) 0.505 □ 0.510
21) 0.772 □ 0.777
22) 0.5 □ 0.500
23) 16.89 □ 15.89
24) 12.25 □ 12.35
25) 5.82 □ 5.69
26) 1.320 □ 1.032
27) 0.082 □ 0.088
28) 0.99 □ 0.099
29) 2.560 □ 1.950
30) 0.770 □ 0.707
31) 15.54 □ 1.554
32) 0.323 □ 0.332

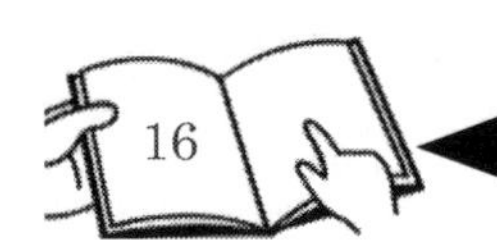

Comparing Decimals – Answers

Compare. Use >, =, and <

1) 0.88 > 0.088
2) 0.56 < 0.57
3) 0.99 > 0.89
4) 1.55 < 1.65
5) 1.58 < 1.75
6) 2.91 > 2.85
7) 14.56 > 1.456
8) 17.85 < 17.89
9) 21.52 > 21.052
10) 11.12 > 11.03
11) 9.650 = 9.65
12) 8.578 > 8.568
13) 3.15 > 0.315
14) 16.61 > 16.16
15) 18.581 > 8.991
16) 25.05 > 2.505
17) 4.55 < 4.65
18) 0.158 < 1.58
19) 0.881 > 0.871
20) 0.505 < 0.510
21) 0.772 < 0.777
22) 0.5 = 0.500
23) 16.89 > 15.89
24) 12.25 < 12.35
25) 5.82 > 5.69
26) 1.320 > 1.032
27) 0.082 < 0.088
28) 0.99 > 0.099
29) 2.560 > 1.950
30) 0.770 > 0.707
31) 15.54 > 1.554
32) 0.323 < 0.332

Rounding Decimals

Round each number to the underlined place value.

1) $\underline{2}.814 =$

2) $3.5\underline{6}2 =$

3) $12.1\underline{2}5 =$

4) $1\underline{5}.5 =$

5) $1.9\underline{8}1 =$

6) $14.\underline{2}15 =$

7) $17.5\underline{4}8 =$

8) $25.5\underline{0}8 =$

9) $3\underline{1}.089 =$

10) $69.\underline{3}45 =$

11) $9.4\underline{5}7 =$

12) $1\underline{2}.901 =$

13) $2.6\underline{5}8 =$

14) $32.\underline{5}65 =$

15) $6.0\underline{5}8 =$

16) $98.1\underline{0}8 =$

17) $27.\underline{7}05 =$

18) $3\underline{6}.75 =$

19) $9.\underline{0}8 =$

20) $7.\underline{1}85 =$

21) $22.5\underline{4}7 =$

22) $66.\underline{0}98 =$

23) $8\underline{7}.75 =$

24) $18.\underline{5}41 =$

25) $10.2\underline{5}8 =$

26) $13.\underline{4}56 =$

27) $71.0\underline{8}4 =$

28) $2\underline{9}.23 =$

29) $45.\underline{5}5 =$

30) $9\underline{1}.08 =$

31) $8\underline{3}.433 =$

32) $74.\underline{6}4 =$

Rounding Decimals - Answers

Round each number to the underlined place value.

1) $\underline{2}.814 = 3$
2) $3.5\underline{6}2 = 3.56$
3) $12.1\underline{2}5 = 12.13$
4) $1\underline{5}.5 = 16$
5) $1.9\underline{8}1 = 1.98$
6) $14.\underline{2}15 = 14.2$
7) $17.5\underline{4}8 = 17.55$
8) $25.5\underline{0}8 = 25.51$
9) $3\underline{1}.089 = 31$
10) $69.\underline{3}45 = 69.3$
11) $9.4\underline{5}7 = 9.46$
12) $1\underline{2}.901 = 13$
13) $2.6\underline{5}8 = 2.66$
14) $32.\underline{5}65 = 32.6$
15) $6.0\underline{5}8 = 6.06$
16) $98.1\underline{0}8 = 98.11$
17) $27.\underline{7}05 = 27.7$
18) $3\underline{6}.75 = 37$
19) $9.\underline{0}8 = 9.1$
20) $7.\underline{1}85 = 7.2$
21) $22.5\underline{4}7 = 22.55$
22) $66.\underline{0}98 = 66.1$
23) $8\underline{7}.75 = 88$
24) $18.\underline{5}41 = 18.5$
25) $10.2\underline{5}8 = 10.26$
26) $13.\underline{4}56 = 13.5$
27) $71.0\underline{8}4 = 71.08$
28) $2\underline{9}.23 = 29$
29) $45.\underline{5}5 = 45.6$
30) $9\underline{1}.08 = 91$
31) $8\underline{3}.433 = 83$
32) $74.\underline{6}4 = 74.6$

Find more at

bit.ly/3mKEluf

Adding and Subtracting Decimals

Solve.

1) $15.63 + 19.64 =$

2) $16.38 + 17.59 =$

3) $75.31 - 59.69 =$

4) $49.38 - 29.89 =$

5) $24.32 + 26.45 =$

6) $36.25 + 18.37 =$

7) $47.85 - 35.12 =$

8) $85.65 - 67.48 =$

9) $25.49 + 34.18 =$

10) $19.99 + 48.66 =$

11) $46.32 - 27.77 =$

12) $54.62 - 48.12 =$

13) $24.42 + 16.54 =$

14) $52.13 + 12.32 =$

15) $82.36 - 78.65 =$

16) $64.12 - 49.15 =$

17) $36.41 + 24.52 =$

18) $85.96 - 74.63 =$

19) $52.62 - 42.54 =$

20) $21.20 + 24.58 =$

21) $32.15 + 17.17 =$

22) $96.32 - 85.54 =$

23) $89.78 - 69.85 =$

24) $29.28 + 39.79 =$

25) $11.11 + 19.99 =$

26) $28.82 + 20.88 =$

27) $63.14 - 28.91 =$

28) $56.61 - 49.72 =$

29) $26.13 + 31.13 =$

30) $30.19 + 20.87 =$

31) $66.24 - 59.10 =$

32) $89.31 - 72.17 =$

Adding and Subtracting Decimals - Answers

Solve.

1) $15.63 + 19.64 = 35.27$
2) $16.38 + 17.59 = 33.97$
3) $75.31 - 59.69 = 15.62$
4) $49.38 - 29.89 = 19.49$
5) $24.32 + 26.45 = 50.77$
6) $36.25 + 18.37 = 54.62$
7) $47.85 - 35.12 = 12.73$
8) $85.65 - 67.48 = 18.17$
9) $25.49 + 34.18 = 59.67$
10) $19.99 + 48.66 = 68.65$
11) $46.32 - 27.77 = 18.55$
12) $54.62 - 48.12 = 6.5$
13) $24.42 + 16.54 = 40.96$
14) $52.13 + 12.32 = 64.45$
15) $82.36 - 78.65 = 3.71$
16) $64.12 - 49.15 = 14.97$
17) $36.41 + 24.52 = 60.93$
18) $85.96 - 74.63 = 11.33$
19) $52.62 - 42.54 = 10.08$
20) $21.20 + 24.58 = 45.78$
21) $32.15 + 17.17 = 49.32$
22) $96.32 - 85.54 = 10.78$
23) $89.78 - 69.85 = 19.93$
24) $29.28 + 39.79 = 69.07$
25) $11.11 + 19.99 = 31.1$
26) $28.82 + 20.88 = 49.7$
27) $63.14 - 28.91 = 34.23$
28) $56.61 - 49.72 = 6.89$
29) $26.13 + 31.13 = 57.26$
30) $30.19 + 20.87 = 51.06$
31) $66.24 - 59.10 = 7.14$
32) $89.31 - 72.17 = 17.14$

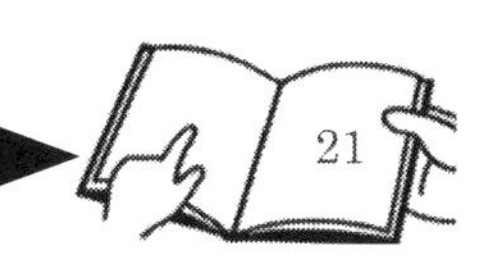

Multiplying and Dividing Decimals

Solve.

1) $11.2 \times 0.4 =$

2) $13.5 \times 0.8 =$

3) $42.2 \div 2 =$

4) $54.6 \div 6 =$

5) $23.1 \times 0.3 =$

6) $1.2 \times 0.7 =$

7) $5.5 \div 0.5 =$

8) $64.8 \div 8 =$

9) $1.4 \times 0.5 =$

10) $4.5 \times 0.3 =$

11) $88.8 \div 4 =$

12) $10.5 \div 5 =$

13) $2.2 \times 0.3 =$

14) $0.2 \times 0.52 =$

15) $95.7 \div 100 =$

16) $36.6 \div 6 =$

17) $3.2 \times 2 =$

18) $4.1 \times 0.5 =$

19) $68.4 \div 2 =$

20) $27.9 \div 9 =$

21) $3.5 \times 4 =$

22) $4.8 \times 0.5 =$

23) $6.4 \div 4 =$

24) $72.8 \div 0.8 =$

25) $1.8 \times 3 =$

26) $6.5 \times 0.2 =$

27) $93.6 \div 3 =$

28) $45.15 \div 0.5 =$

29) $13.2 \times 0.4 =$

30) $11.2 \times 5 =$

31) $7.2 \div 0.8 =$

32) $96.4 \div 0.2 =$

Multiplying and Dividing Decimals – Answers

Solve.

1) $11.2 \times 0.4 = 4.48$
2) $13.5 \times 0.8 = 10.8$
3) $42.2 \div 2 = 21.1$
4) $54.6 \div 6 = 9.1$
5) $23.1 \times 0.3 = 6.93$
6) $1.2 \times 0.7 = 0.84$
7) $5.5 \div 5 = 1.1$
8) $64.8 \div 8 = 8.1$
9) $1.4 \times 0.5 = 0.7$
10) $4.5 \times 0.3 = 1.35$
11) $88.8 \div 4 = 22.2$
12) $10.5 \div 5 = 2.1$
13) $2.2 \times 0.3 = 0.66$
14) $0.2 \times 0.52 = 0.104$
15) $95.7 \div 100 = 0.957$
16) $36.6 \div 6 = 6.1$
17) $3.2 \times 2 = 6.4$
18) $4.1 \times 0.5 = 2.05$
19) $68.4 \div 2 = 34.2$
20) $27.9 \div 9 = 3.1$
21) $3.5 \times 4 = 14$
22) $4.8 \times 0.5 = 2.4$
23) $6.4 \div 4 = 1.6$
24) $72.8 \div 0.8 = 91$
25) $1.8 \times 3 = 5.4$
26) $6.5 \times 0.2 = 1.3$
27) $93.6 \div 3 = 31.2$
28) $45.15 \div 0.5 = 90.3$
29) $13.2 \times 0.4 = 5.28$
30) $11.2 \times 5 = 56$
31) $7.2 \div 0.8 = 9$
32) $96.4 \div 0.2 = 482$

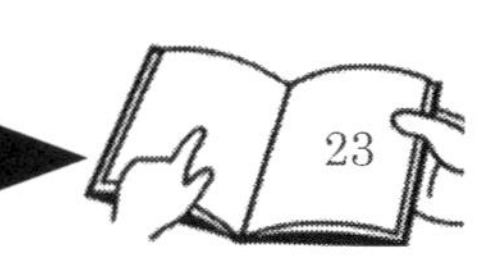

Adding and Subtracting Integers

Solve.

1) $-(8) + 13 =$

2) $17 - (-12 - 8) =$

3) $(-15) + (-4) =$

4) $(-14) + (-8) + 9 =$

5) $-(23) + 19 =$

6) $(-7 + 5) - 9 =$

7) $28 + (-32) =$

8) $(-11) + (-9) + 5 =$

9) $25 - (8 - 7) =$

10) $-(29) + 17 =$

11) $(-38) + (-3) + 29 =$

12) $15 - (-7 + 9) =$

13) $24 - (8 - 2) =$

14) $(-7 + 4) - 9 =$

15) $(-17) + (-3) + 9 =$

16) $(-26) + (-7) + 8 =$

17) $(-9) + (-11) =$

18) $8 - (-23 - 13) =$

19) $(-16) + (-2) =$

20) $25 - (7 - 4) =$

21) $23 + (-12) =$

22) $(-18) + (-6) =$

23) $17 - (-21 - 7) =$

24) $-(28) - (-16) + 5 =$

25) $(-9 + 4) - 8 =$

26) $(-28) + (-6) + 17 =$

27) $-(21) - (-15) + 9 =$

28) $(-31) + (-6) =$

29) $(-17) + (-11) + 14 =$

30) $(-29) + (-10) + 13 =$

31) $-(24) - (-12) + 5 =$

32) $8 - (-19 - 10) =$

Adding and Subtracting Integers – Answers

Solve.

1) $-(8) + 13 = 5$
2) $17 - (-12 - 8) = 37$
3) $(-15) + (-4) = -19$
4) $(-14) + (-8) + 9 = -13$
5) $-(23) + 19 = -4$
6) $(-7 + 5) - 9 = -11$
7) $28 + (-32) = -4$
8) $(-11) + (-9) + 5 = -15$
9) $25 - (8 - 7) = 24$
10) $-(29) + 17 = -12$
11) $(-38) + (-3) + 29 = -12$
12) $15 - (-7 + 9) = 13$
13) $24 - (8 - 2) = 18$
14) $(-7 + 4) - 9 = -12$
15) $(-17) + (-3) + 9 = -11$
16) $(-26) + (-7) + 8 = -25$
17) $(-9) + (-11) = -20$
18) $8 - (-23 - 13) = 44$
19) $(-16) + (-2) = -18$
20) $25 - (7 - 4) = 22$
21) $23 + (-12) = 11$
22) $(-18) + (-6) = -24$
23) $17 - (-21 - 7) = 45$
24) $-(28) - (-16) + 5 = -7$
25) $(-9 + 4) - 8 = -13$
26) $(-28) + (-6) + 17 = -17$
27) $-(21) - (-15) + 9 = 3$
28) $(-31) + (-6) = -37$
29) $(-17) + (-11) + 14 = -14$
30) $(-29) + (-10) + 13 = -26$
31) $-(24) - (-12) + 5 = -7$
32) $8 - (-19 - 10) = 37$

Multiplying and Dividing Integers

Solve.

1) $(-9) \times (-8) =$

2) $6 \times (-6) =$

3) $49 \div (-7) =$

4) $(-64) \div 8 =$

5) $(4) \times (-6) =$

6) $(-9) \times (-11) =$

7) $(10) \div (-5) =$

8) $144 \div (-12) =$

9) $(10) \times (-2) =$

10) $(-8) \times (-2) \times 5 =$

11) $(8) \div (-2) =$

12) $45 \div (-15) =$

13) $(5) \times (-7) =$

14) $(-6) \times (-5) \times 5 =$

15) $(12) \div (-6) =$

16) $(14) \div (-7) =$

17) $196 \div (-14) =$

18) $(27 - 13) \times (-2) =$

19) $125 \div (-5) =$

20) $66 \div (-6) =$

21) $(-6) \times (-5) \times 3 =$

22) $(15 - 6) \times (-3) =$

23) $(32 - 24) \div (-4) =$

24) $72 \div (-6) =$

25) $(-14 + 8) \times (-7) =$

26) $(-3) \times (-9) \times 3 =$

27) $84 \div (-12) =$

28) $(-12) \times (-10) =$

29) $25 \times (-4) =$

30) $(-3) \times (-5) \times 5 =$

31) $(15) \div (-3) =$

32) $(-18) \div (3) =$

Multiplying and Dividing Integers - Answers

Solve.

1) $(-9) \times (-8) = 72$
2) $6 \times (-6) = -36$
3) $49 \div (-7) = -7$
4) $(-64) \div 8 = -8$
5) $(4) \times (-6) = -24$
6) $(-9) \times (-11) = 99$
7) $(10) \div (-5) = -2$
8) $144 \div (-12) = -12$
9) $(10) \times (-2) = -20$
10) $(-8) \times (-2) \times 5 = 80$
11) $(8) \div (-2) = -4$
12) $45 \div (-15) = -3$
13) $(5) \times (-7) = -35$
14) $(-6) \times (-5) \times 5 = 150$
15) $(12) \div (-6) = -2$
16) $(14) \div (-7) = -2$
17) $196 \div (-14) = -14$
18) $(27 - 13) \times (-2) = -28$
19) $125 \div (-5) = -25$
20) $66 \div (-6) = -11$
21) $(-6) \times (-5) \times 3 = 90$
22) $(15 - 6) \times (-3) = -27$
23) $(32 - 24) \div (-4) = -2$
24) $72 \div (-6) = -12$
25) $(-14 + 8) \times (-7) = 42$
26) $(-3) \times (-9) \times 3 = 81$
27) $84 \div (-12) = -7$
28) $(-12) \times (-10) = 120$
29) $25 \times (-4) = -100$
30) $(-3) \times (-5) \times 5 = 75$
31) $(15) \div (-3) = -5$
32) $(-18) \div (3) = -6$

bit.ly/3pjQW98

Order of Operation

Calculate.

1) $18 + (32 \div 4) =$
2) $(3 \times 8) \div (-2) =$
3) $67 - (4 \times 8) =$
4) $(-11) \times (8 - 3) =$
5) $(18 - 7) \times (6) =$
6) $(6 \times 10) \div (12 + 3) =$
7) $(13 \times 2) - (24 \div 6) =$
8) $(-5) + (4 \times 3) + 8 =$
9) $(4 \times 2^3) + (16 - 9) =$
10) $(3^2 \times 7) \div (-2 + 1) =$
11) $[-2(48 \div 2^3)] - 6 =$
12) $(-4) + (7 \times 8) + 18 =$
13) $(3 \times 7) + (16 - 7) =$
14) $[3^3 \times (48 \div 2^3)] \div (-2) =$
15) $(14 \times 3) - (3^4 \div 9) =$
16) $(96 \div 12) \times (-3) =$
17) $(48 \div 2^2) \times (-2) =$
18) $(56 \div 7) \times (-5) =$
19) $(-2^2) + (7 \times 9) - 21 =$
20) $(2^4 - 9) \times (-6) =$
21) $[4^3 \times (50 \div 5^2)] \div (-16) =$
22) $(3^2 \times 4^2) \div (-4 + 2) =$
23) $6^2 - (-6 \times 4) + 3 =$
24) $4^2 - (5^2 \times 3) =$
25) $(-4) + (12^2 \div 3^2) - 7^2 =$
26) $(3^2 \times 5) + (-5^2 - 9) =$
27) $2[(3^2 \times 5) \times (-6)] =$
28) $(11^2 - 2^2) - (-7^2) =$
29) $(2^3 \times 3) - (49 \div 7) =$
30) $3[(3^2 \times 5) + (25 \div 5)] =$
31) $(6^2 \times 5) \div (-5) =$
32) $2^2[(6^3 \div 12) - (3^4 \div 27)] =$

Order of Operation – Answers

Calculate.

1) $18 + (32 \div 4) = 26$
2) $(3 \times 8) \div (-2) = -12$
3) $67 - (4 \times 8) = 35$
4) $(-11) \times (8 - 3) = -55$
5) $(18 - 7) \times (6) = 66$
6) $(6 \times 10) \div (12 + 3) = 4$
7) $(13 \times 2) - (24 \div 6) = 22$
8) $(-5) + (4 \times 3) + 8 = 15$
9) $(4 \times 2^3) + (16 - 9) = 39$
10) $(3^2 \times 7) \div (-2 + 1) = -63$
11) $[-2(48 \div 2^3)] - 6 = -18$
12) $(-4) + (7 \times 8) + 18 = 70$
13) $(3 \times 7) + (16 - 7) = 30$
14) $[3^3 \times (48 \div 2^3)] \div (-2) = -81$
15) $(14 \times 3) - (3^4 \div 9) = 33$
16) $(96 \div 12) \times (-3) = -24$
17) $(48 \div 2^2) \times (-2) = -24$
18) $(56 \div 7) \times (-5) = -40$
19) $(-2^2) + (7 \times 9) - 21 = 38$
20) $(2^4 - 9) \times (-6) = -42$
21) $[4^3 \times (50 \div 5^2)] \div (-16) = -8$
22) $(3^2 \times 4^2) \div (-4 + 2) = -72$
23) $6^2 - (-6 \times 4) + 3 = 63$
24) $4^2 - (5^2 \times 3) = -59$
25) $(-4) + (12^2 \div 3^2) - 7^2 = -37$
26) $(3^2 \times 5) + (-5^2 - 9) = 11$
27) $2[(3^2 \times 5) \times (-6)] = -540$
28) $(11^2 - 2^2) - (-7^2) = 166$
29) $(2^3 \times 3) - (49 \div 7) = 17$
30) $3[(3^2 \times 5) + (25 \div 5)] = 150$
31) $(6^2 \times 5) \div (-5) = -36$
32) $2^2[(6^3 \div 12) - (3^4 \div 27)] = 60$

Integers and Absolute Value

Calculate.

1) $5 - |8 - 12| =$

2) $|15| - \frac{|-16|}{4} =$

3) $\frac{|9 \times -6|}{18} \times \frac{|-24|}{8} =$

4) $|13 \times 3| + \frac{|-72|}{9} =$

5) $4 - |11 - 18| - |3| =$

6) $|18| - \frac{|-12|}{4} =$

7) $\frac{|5 \times -8|}{10} \times \frac{|-22|}{11} =$

8) $|9 \times 3| + \frac{|-36|}{4} =$

9) $|-42 + 7| \times \frac{|-2 \times 5|}{10} =$

10) $6 - |17 - 11| - |5| =$

11) $|13| - \frac{|-54|}{6} =$

12) $\frac{|9 \times -4|}{12} \times \frac{|-45|}{9} =$

13) $|-75 + 50| \times \frac{|-4 \times 5|}{5} =$

14) $\frac{|-26|}{13} \times \frac{|-32|}{8} =$

15) $14 - |8 - 18| - |-12| =$

16) $|29| - \frac{|-20|}{5} =$

17) $\frac{|3 \times 8|}{2} \times \frac{|-33|}{3} =$

18) $|-45 + 15| \times \frac{|-12 \times 5|}{6} =$

19) $\frac{|-50|}{5} \times \frac{|-77|}{11} =$

20) $12 - |2 - 7| - |15| =$

21) $|18| - \frac{|-45|}{15} =$

22) $\frac{|7 \times 8|}{4} \times \frac{|-48|}{12} =$

23) $\frac{|30 \times 2|}{3} \times |-12| =$

24) $\frac{|-36|}{9} \times \frac{|-80|}{8} =$

25) $|-35 + 8| \times \frac{|-9 \times 5|}{15} =$

26) $|19| - \frac{|-18|}{2} =$

27) $14 - |11 - 23| + |2| =$

28) $|-39 + 7| \times \frac{|-4 \times 6|}{3} =$

Integers and Absolute Value - Answers

Calculate.

1) $5 - |8 - 12| = 1$
2) $|15| - \frac{|-16|}{4} = 11$
3) $\frac{|9 \times -6|}{18} \times \frac{|-24|}{8} = 9$
4) $|13 \times 3| + \frac{|-72|}{9} = 47$
5) $4 - |11 - 18| - |3| = -6$
6) $|18| - \frac{|-12|}{4} = 15$
7) $\frac{|5 \times -8|}{10} \times \frac{|-22|}{11} = 8$
8) $|9 \times 3| + \frac{|-36|}{4} = 36$
9) $|-42 + 7| \times \frac{|-2\times5|}{10} = 35$
10) $6 - |17 - 11| - |5| = -5$
11) $|13| - \frac{|-54|}{6} = 4$
12) $\frac{|9 \times -4|}{12} \times \frac{|-45|}{9} = 15$
13) $|-75 + 50| \times \frac{|-4\times5|}{5} = 100$
14) $\frac{|-26|}{13} \times \frac{|-32|}{8} = 8$
15) $14 - |8 - 18| - |-12| = -8$
16) $|29| - \frac{|-20|}{5} = 25$
17) $\frac{|3\times 8|}{2} \times \frac{|-33|}{3} = 132$
18) $|-45 + 15| \times \frac{|-12\times5|}{6} = 300$
19) $\frac{|-50|}{5} \times \frac{|-77|}{11} = 70$
20) $12 - |2 - 7| - |15| = -8$
21) $|18| - \frac{|-45|}{15} = 15$
22) $\frac{|7 \times 8|}{4} \times \frac{|-48|}{12} = 56$
23) $\frac{|30 \times 2|}{3} \times |-12| = 240$
24) $\frac{|-36|}{9} \times \frac{|-80|}{8} = 40$
25) $|-35 + 8| \times \frac{|-9\times5|}{15} = 81$
26) $|19| - \frac{|-18|}{2} = 10$
27) $14 - |11 - 23| + |2| = 4$
28) $|-39 + 7| \times \frac{|-4\times6|}{3} = 256$

Simplifying Ratios

Simplify each ratio.

1) $3:27 =$ ___:___

2) $2:8 =$ ___:___

3) $\frac{4}{28} = -$

4) $\frac{16}{40} = -$

5) $10:30 =$ ___:___

6) $5:30 =$ ___:___

7) $\frac{34}{38} = -$

8) $\frac{45}{63} = -$

9) $10:45 =$ ___:___

10) $20:30 =$ ___:___

11) $\frac{40}{64} = -$

12) $\frac{10}{110} = -$

13) $8:12 =$ ___:___

14) $16:20 =$ ___:___

15) $\frac{24}{48} = -$

16) $\frac{21}{77} = -$

17) $8:24 =$ ___:___

18) 9 to $36 =$ ___:___

19) $\frac{64}{72} = -$

20) $\frac{45}{60} = -$

21) $12:15 =$ ___:___

22) $18:54 =$ ___:___

23) $\frac{36}{54} = -$

24) $\frac{48}{104} = -$

25) $15:75 =$ ___:___

26) $16:48 =$ ___:___

27) $\frac{15}{65} = -$

28) $\frac{44}{52} = -$

Simplifying Ratios – Answers

Simplify each ratio.

1) $3:27 = 1:9$

2) $2:8 = 1:4$

3) $\frac{4}{28} = \frac{1}{7}$

4) $\frac{16}{40} = \frac{2}{5}$

5) $10:30 = 1:3$

6) $5:30 = 1:6$

7) $\frac{34}{38} = \frac{17}{19}$

8) $\frac{45}{63} = \frac{5}{7}$

9) $10:45 = 2:9$

10) $20:30 = 2:3$

11) $\frac{40}{64} = \frac{5}{8}$

12) $\frac{10}{110} = \frac{1}{11}$

13) $8:12 = 2:3$

14) $16:20 = 4:5$

15) $\frac{24}{48} = \frac{1}{2}$

16) $\frac{21}{77} = \frac{3}{11}$

17) $8:24 = 1:3$

18) 9 to 36 = 1 to 4

19) $\frac{64}{72} = \frac{8}{9}$

20) $\frac{45}{60} = \frac{3}{4}$

21) $12:15 = 4:5$

22) $18:54 = 1:3$

23) $\frac{36}{54} = \frac{2}{3}$

24) $\frac{48}{104} = \frac{6}{13}$

25) $15:75 = 1:5$

26) $16:48 = 1:3$

27) $\frac{15}{65} = \frac{3}{13}$

28) $\frac{44}{52} = \frac{11}{13}$

Proportional Ratios

Solve each proportion for x.

1) $\frac{4}{7} = \frac{16}{x}, x =$ ____

2) $\frac{4}{9} = \frac{x}{18}, x =$ ____

3) $\frac{3}{5} = \frac{24}{x}, x =$ ____

4) $\frac{3}{10} = \frac{x}{50}, x =$ ____

5) $\frac{3}{11} = \frac{15}{x}, x =$ ____

6) $\frac{6}{15} = \frac{x}{45}, x =$ ____

7) $\frac{6}{19} = \frac{12}{x}, x =$ ____

8) $\frac{7}{16} = \frac{x}{32}, x =$ ____

9) $\frac{18}{21} = \frac{54}{x}, x =$ ____

10) $\frac{13}{15} = \frac{39}{x}, x =$ ____

11) $\frac{9}{13} = \frac{72}{x}, x =$ ____

12) $\frac{8}{30} = \frac{x}{180}, x =$ ____

13) $\frac{3}{19} = \frac{9}{x}, x =$ ____

14) $\frac{1}{3} = \frac{x}{90}, x =$ ____

15) $\frac{25}{45} = \frac{x}{9}, x =$ ____

16) $\frac{1}{6} = \frac{9}{x}, x =$ ____

17) $\frac{7}{9} = \frac{63}{x}, x =$ ____

18) $\frac{54}{72} = \frac{x}{8}, x =$ ____

19) $\frac{32}{40} = \frac{4}{x}, x =$ ____

20) $\frac{21}{42} = \frac{x}{6}, x =$ ____

21) $\frac{56}{72} = \frac{7}{x}, x =$ ____

22) $\frac{1}{14} = \frac{x}{42}, x =$ ____

23) $\frac{5}{7} = \frac{75}{x}, x =$ ____

24) $\frac{30}{48} = \frac{x}{8}, x =$ ____

25) $\frac{36}{88} = \frac{9}{x}, x =$ ____

26) $\frac{62}{68} = \frac{x}{34}, x =$ ____

27) $\frac{42}{60} = \frac{x}{10}, x =$ ____

28) $\frac{8}{9} = \frac{x}{108}, x =$ ____

29) $\frac{46}{69} = \frac{x}{3}, x =$ ____

30) $\frac{99}{121} = \frac{x}{11}, x =$ ____

31) $\frac{19}{21} = \frac{x}{63}, x =$ ____

32) $\frac{11}{12} = \frac{x}{48}, x =$ ____

Proportional Ratios - Answers

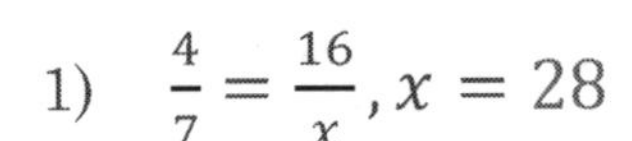
Solve each proportion for x.

1) $\frac{4}{7}=\frac{16}{x}, x=28$
2) $\frac{4}{9}=\frac{x}{18}, x=8$
3) $\frac{3}{5}=\frac{24}{x}, x=40$
4) $\frac{3}{10}=\frac{x}{50}, x=15$
5) $\frac{3}{11}=\frac{15}{x}, x=55$
6) $\frac{6}{15}=\frac{x}{45}, x=18$
7) $\frac{6}{19}=\frac{12}{x}, x=38$
8) $\frac{7}{16}=\frac{x}{32}, x=14$
9) $\frac{18}{21}=\frac{54}{x}, x=63$
10) $\frac{13}{15}=\frac{39}{x}, x=45$
11) $\frac{9}{13}=\frac{72}{x}, x=104$
12) $\frac{8}{30}=\frac{x}{180}, x=48$
13) $\frac{3}{19}=\frac{9}{x}, x=57$
14) $\frac{1}{3}=\frac{x}{90}, x=30$
15) $\frac{25}{45}=\frac{x}{9}, x=5$
16) $\frac{1}{6}=\frac{9}{x}, x=54$
17) $\frac{7}{9}=\frac{63}{x}, x=81$
18) $\frac{54}{72}=\frac{x}{8}, x=6$
19) $\frac{32}{40}=\frac{4}{x}, x=5$
20) $\frac{21}{42}=\frac{x}{6}, x=3$
21) $\frac{56}{72}=\frac{7}{x}, x=9$
22) $\frac{1}{14}=\frac{x}{42}, x=3$
23) $\frac{5}{7}=\frac{75}{x}, x=105$
24) $\frac{30}{48}=\frac{x}{8}, x=5$
25) $\frac{36}{88}=\frac{9}{x}, x=22$
26) $\frac{62}{68}=\frac{x}{34}, x=31$
27) $\frac{42}{60}=\frac{x}{10}, x=7$
28) $\frac{8}{9}=\frac{x}{108}, x=96$
29) $\frac{46}{69}=\frac{x}{3}, x=2$
30) $\frac{99}{121}=\frac{x}{11}, x=9$
31) $\frac{19}{21}=\frac{x}{63}, x=57$
32) $\frac{11}{12}=\frac{x}{48}, x=44$

Create Proportion

State if each pair of ratios form a proportion.

1) $\frac{5}{8}$ and $\frac{25}{50}$

2) $\frac{2}{11}$ and $\frac{4}{22}$

3) $\frac{2}{5}$ and $\frac{8}{20}$

4) $\frac{3}{11}$ and $\frac{9}{33}$

5) $\frac{5}{10}$ and $\frac{15}{30}$

6) $\frac{4}{13}$ and $\frac{8}{24}$

7) $\frac{6}{9}$ and $\frac{24}{36}$

8) $\frac{7}{12}$ and $\frac{14}{20}$

9) $\frac{3}{8}$ and $\frac{27}{72}$

10) $\frac{12}{20}$ and $\frac{36}{60}$

11) $\frac{11}{12}$ and $\frac{55}{60}$

12) $\frac{12}{15}$ and $\frac{24}{25}$

13) $\frac{15}{19}$ and $\frac{20}{38}$

14) $\frac{10}{14}$ and $\frac{40}{56}$

15) $\frac{11}{13}$ and $\frac{44}{39}$

16) $\frac{15}{16}$ and $\frac{30}{32}$

17) $\frac{17}{19}$ and $\frac{34}{48}$

18) $\frac{5}{18}$ and $\frac{15}{54}$

19) $\frac{3}{14}$ and $\frac{18}{42}$

20) $\frac{7}{11}$ and $\frac{14}{32}$

21) $\frac{8}{11}$ and $\frac{32}{44}$

22) $\frac{9}{13}$ and $\frac{18}{26}$

Solve.

23) The ratio of boys to girls in a class is 5:6. If there are 25 boys in the class, how many girls are in that class? _____

24) The ratio of red marbles to blue marbles in a bag is 4:7. If there are 77 marbles in the bag, how many of the marbles are red? ______________

25) You can buy 8 cans of green beans at a supermarket for $3.20. How much does it cost to buy 48 cans of green beans? ______________

Create Proportion – Answers

State if each pair of ratios form a proportion.

1) $\frac{5}{8}$ *and* $\frac{25}{50}$, *No*
2) $\frac{2}{11}$ *and* $\frac{4}{22}$, *Yes*
3) $\frac{2}{5}$ *and* $\frac{8}{20}$, *Yes*
4) $\frac{3}{11}$ *and* $\frac{9}{33}$, *Yes*
5) $\frac{5}{10}$ *and* $\frac{15}{30}$, *Yes*
6) $\frac{4}{13}$ *and* $\frac{8}{24}$, *No*
7) $\frac{6}{9}$ *and* $\frac{24}{36}$, *Yes*
8) $\frac{7}{12}$ *and* $\frac{14}{20}$, *No*
9) $\frac{3}{8}$ *and* $\frac{27}{72}$, *Yes*
10) $\frac{12}{20}$ *and* $\frac{36}{60}$, *Yes*
11) $\frac{11}{12}$ *and* $\frac{55}{60}$, *Yes*
12) $\frac{12}{15}$ *and* $\frac{24}{25}$, *No*
13) $\frac{15}{19}$ *and* $\frac{20}{38}$, *No*
14) $\frac{10}{14}$ *and* $\frac{40}{56}$, *Yes*
15) $\frac{11}{13}$ *and* $\frac{44}{39}$, *No*
16) $\frac{15}{16}$ *and* $\frac{30}{32}$, *Yes*
17) $\frac{17}{19}$ *and* $\frac{34}{38}$, *Yes*
18) $\frac{5}{18}$ *and* $\frac{15}{54}$, *Yes*
19) $\frac{3}{14}$ *and* $\frac{18}{42}$, *No*
20) $\frac{7}{11}$ *and* $\frac{14}{32}$, *No*
21) $\frac{8}{11}$ *and* $\frac{32}{44}$, *Yes*
22) $\frac{9}{13}$ *and* $\frac{18}{26}$, *Yes*

Solve.

23) The ratio of boys to girls in a class is $5:6$. If there are 25 boys in the class, how many girls are in that class? **30 girls**

24) The ratio of red marbles to blue marbles in a bag is $4:7$. If there are 77 marbles in the bag, how many of the marbles are red? **28 red marbles**

25) You can buy 8 cans of green beans at a supermarket for \$3.20. How much does it cost to buy 48 cans of green beans? $\$\mathbf{19.20}$

Similarity and Ratios

Each pair of figures is similar. Find the missing side.

1)

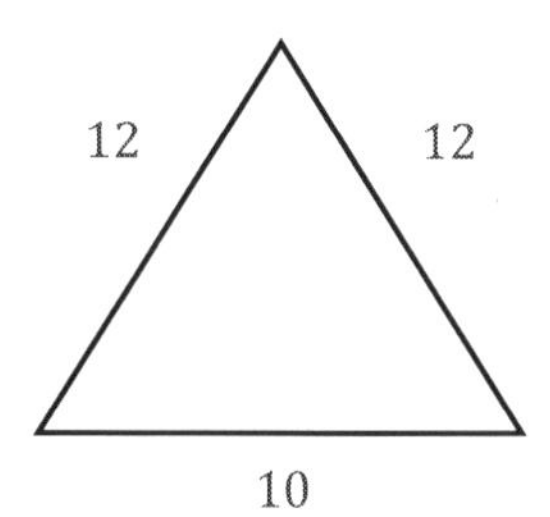

2)

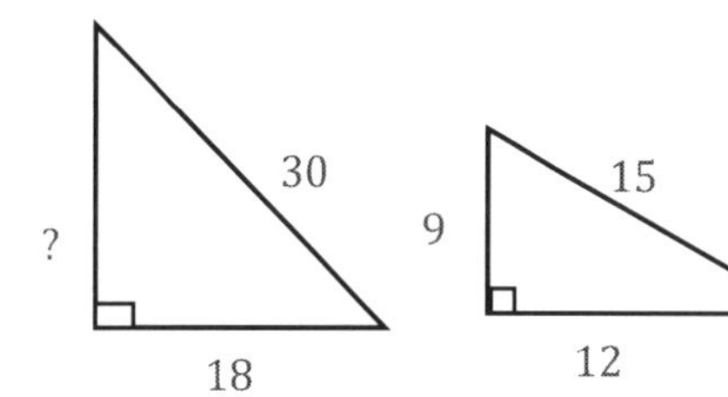

3)

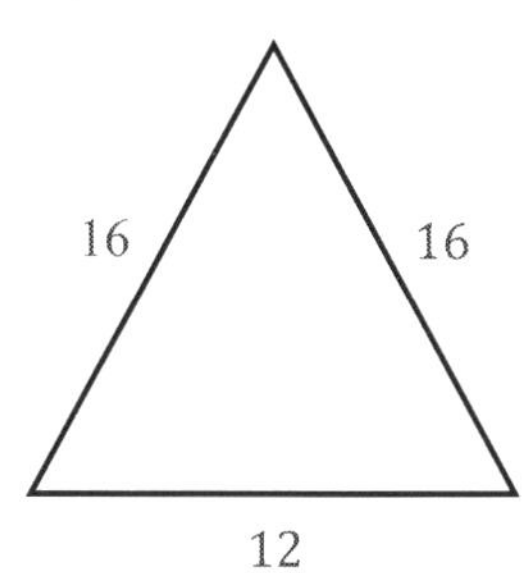

4)

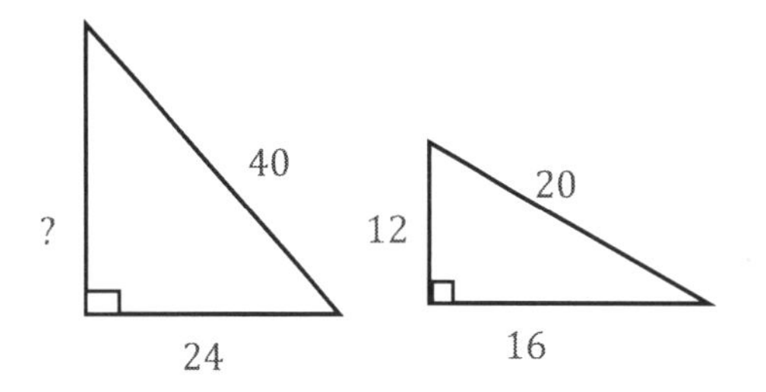

5)

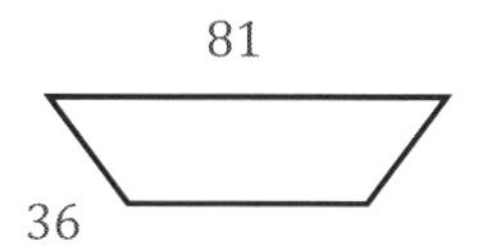

6)

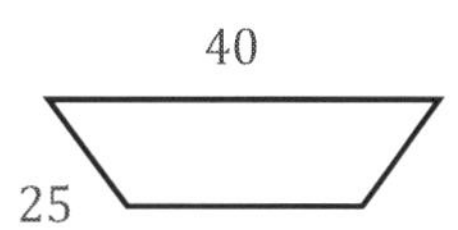

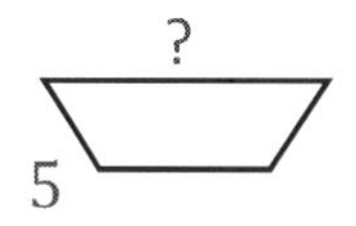

7)

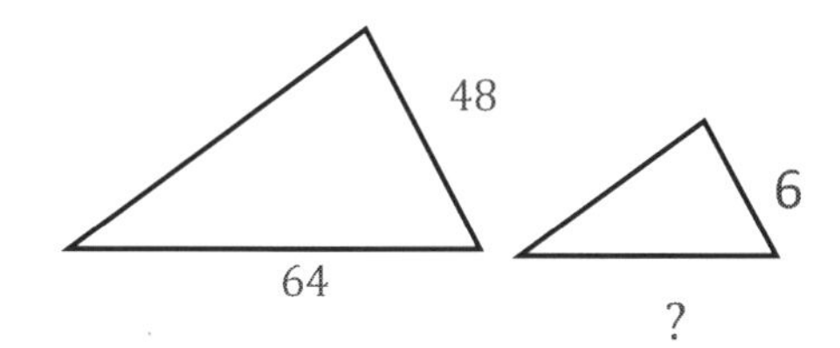

8)

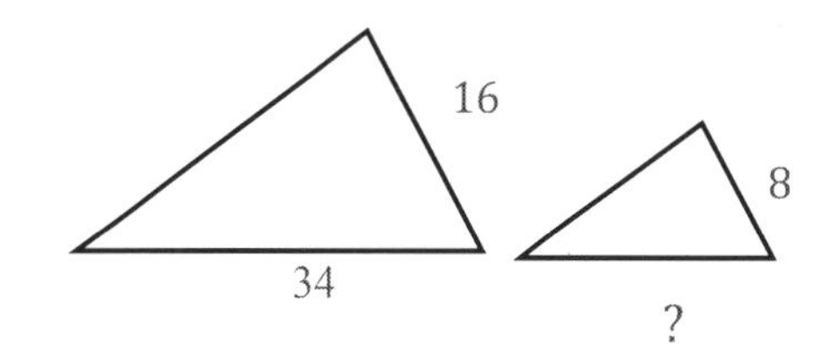

Similarity and Ratios – Answers

 ***Each pair of figures is similar. Find the missing side.***

1) **5**

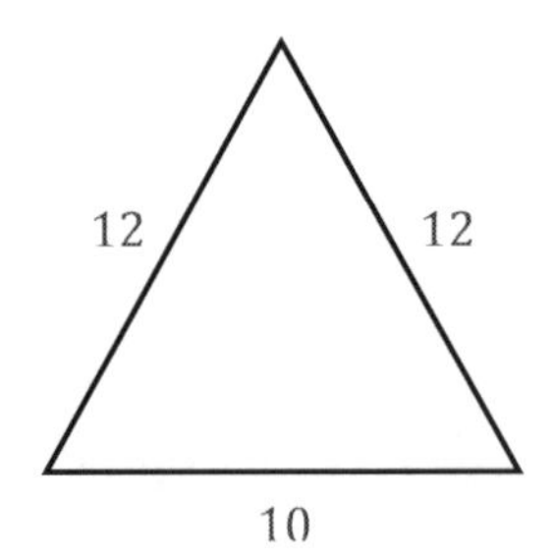

2) **24**

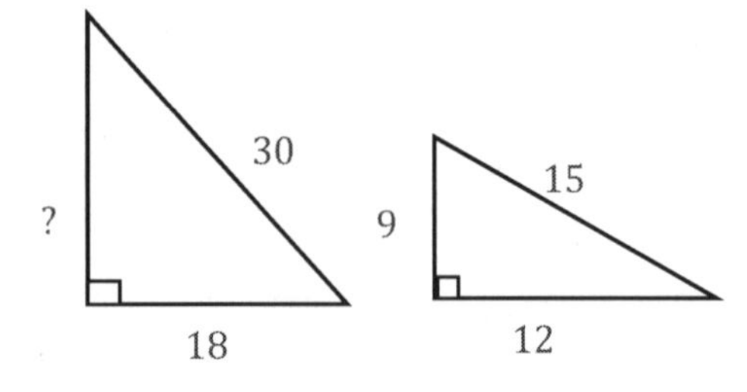

3) **3**

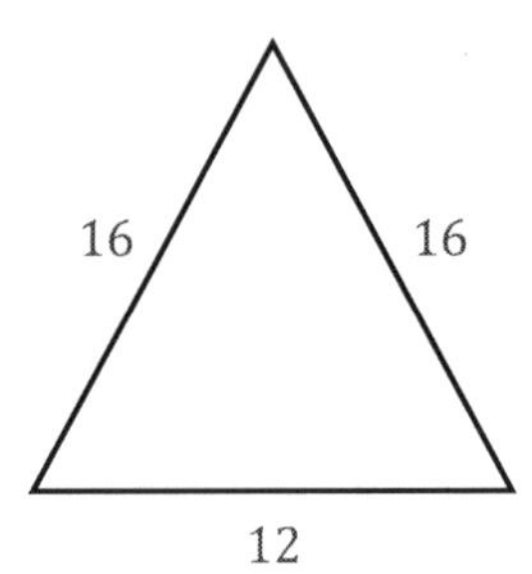

4) 32

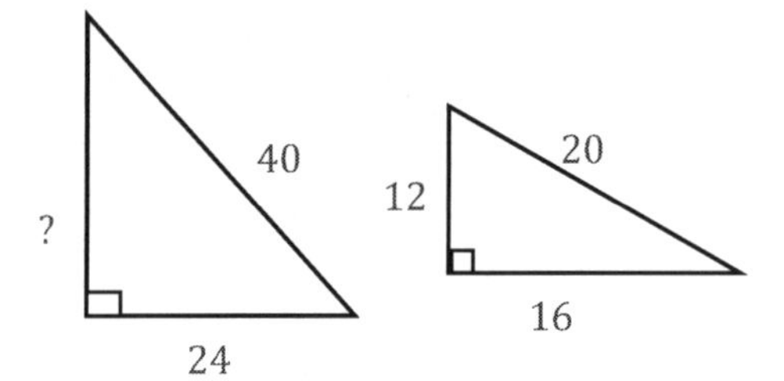

5) **9**

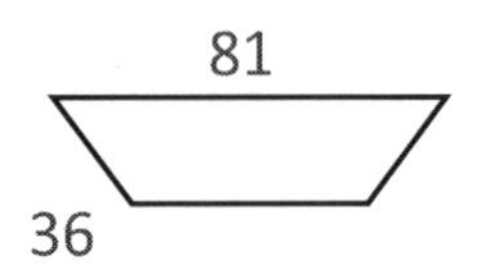

6) 8

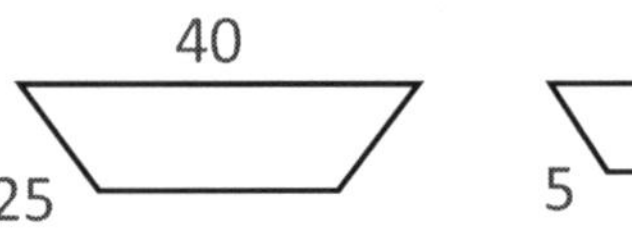

7) **8**

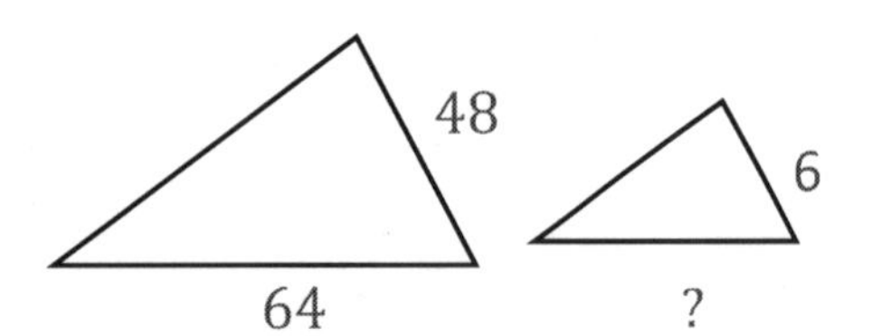

8) 17

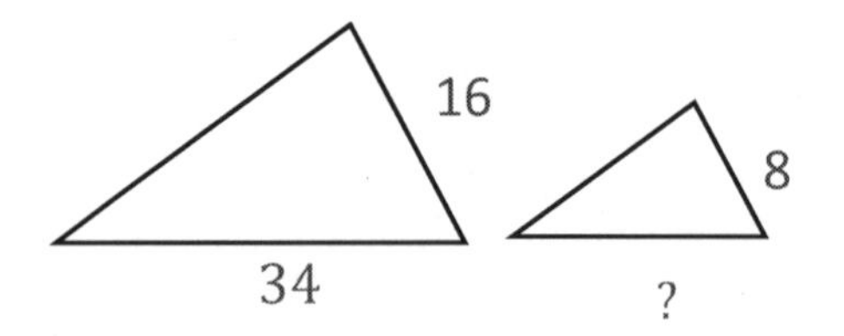

Simple Interest

Determine the simple interest for the following loans.

1) $440 at 5% for 6 years. $___
2) $460 at 2.5% for 4 years. $_
3) $500 at 3% for 5 years. $___
4) $550 at 9% for 2 years. $___
5) $690 at 5% for 6 months. $___
6) $620 at 7% for 3 years. $___
7) $650 at 4.5% for 10 years. $___
8) $850 at 4% for 2 years. $___
9) $640 at 7% for 3 years. $___
10) $300 at 9% for 9 months. $___
11) $760 at 8% for 2 years. $_
12) $910 at 5% for 5 years. $___
13) $540 at 3% for 6 years. $___
14) $780 at 2.5% for 4 years. $___
15) $1,600 at 7% for 3 months. $___
16) $310 at 4% for 4 years. $___
17) $950 at 6% for 5 years. $___
18) $280 at 8% for 7 years. $___
19) $310 at 6% for 3 years. $___
20) $990 at 5% for 4 months. $____
21) $380 at 6% for 5 years. $___
22) $580 at 6% for 4 years. $___
23) $1,200 at 4% for 5 years. $___
24) $3,100 at 5% for 6 years. $___
25) $5,200 at 8% for 2 years. $___
26) $1,400 at 4% for 3 years. $___
27) $300 at 3% for 8 months. $___
28) $150 at 3.5% for 4 years. $___
29) $170 at 6% for 2 years. $___
30) $940 at 8% for 5 years. $___
31) $960 at 1.5% for 8 years. $_
32) $240 at 5% for 4 months. $___
33) $280 at 2% for 5 years. $___
34) $880 at 3% for 2 years. $___
35) $2,200 at 4.5% for 2 years. $___
36) $2,400 at 7% for 3 years. $___
37) $1,800 at 5% for 6 months. $___
38) $190 at 4% for 2 years. $___
39) $560 at 7% for 4 years. $___
40) $720 at 8% for 2 years. $_
41) $780 at 5% for 8 years. $___
42) $880 at 6% for 3 months. $___

Simple Interest - Answers

Determine the simple interest for the following loans.

1) $440 at 5% for 6 years. $132
2) $460 at 2.5% for 4 years. $46
3) $500 at 3% for 5 years. $75
4) $550 at 9% for 2 years. $99
5) $690 at 5% for 6 months. $17.25
6) $620 at 7% for 3 years. $130.20
7) $650 at 4.5% for 10 years. $292.50
8) $850 at 4% for 2 years. $68
9) $640 at 7% for 3 years. $134.40
10) $300 at 9% for 9 months. $20.25
11) $760 at 8% for 2 years. $121.60
12) $910 at 5% for 5 years. $227.50
13) $540 at 3% for 6 years. $97.20
14) $780 at 2.5% for 4 years. $78
15) $1,600 at 7% for 3 months. $28
16) $310 at 4% for 4 years. $49.60
17) $950 at 6% for 5 years. $285
18) $280 at 8% for 7 years. $156.80
19) $310 at 6% for 3 years. $55.80
20) $990 at 5% for 4 months. $198
21) $380 at 6% for 5 years. $114
22) $580 at 6% for 4 years. $139.20
23) $1,200 at 4% for 5 years. $240
24) $3,100 at 5% for 6 years. $930
25) $5,200 at 8% for 2 years. $832
26) $1,400 at 4% for 3 years. $168
27) $300 at 3% for 8 months. $6
28) $150 at 3.5% for 4 years. $21
29) $170 at 6% for 2 years. $16.5
30) $940 at 8% for 5 years. $376
31) $960 at 1.5% for 8 years. $115.20
32) $240 at 5% for 4 months. $4
33) $280 at 2% for 5 years. $28
34) $880 at 3% for 2 years. $52.80
35) $2,200 at 4.5% for 2 years. $198
36) $2,400 at 7% for 3 years. $504
37) $1,800 at 5% for 6 months. $45
38) $190 at 4% for 2 years. $15.20
39) $560 at 7% for 4 years. $156.80
40) $720 at 8% for 2 years. $115.20
41) $780 at 5% for 8 years. $312
42) $880 at 6% for 3 months. $13.20

Percent Problems

Solve each problem.

1) What is 5 percent of 300? ____
2) What is 15 percent of 600? _____
3) What is 12 percent of 450? _____
4) What is 30 percent of 240? _____
5) What is 60 percent of 850? _____
6) 63 is what percent of 300? _____%
7) 80 is what percent of 400? _____%
8) 70 is what percent of 700? _____%
9) 84 is what percent of 600? ____%
10) 90 is what percent of 300? ____%
11) 24 is what percent of 150? ____%
12) 12 is what percent of 80? _____%
13) 4 is what percent of 50? _____%
14) 110 is what percent of 500? __%
15) 16 is what percent of 400? __%
16) 39 is what percent of 300? ____%
17) 56 is what percent of 200? ____%
18) 30 is what percent of 500? ____%
19) 84 is what percent of 700? ____%
20) 40 is what percent of 500? ___%
21) 26 is what percent of 100? ___ %
22) 45 is what percent of 900? ___%
23) 60 is what percent of 400? _____%
24) 18 is what percent of 900? _____%
25) 75 is what percent of 250? _____%
26) 27 is what percent of 900? _____%
27) 49 is what percent of 700? _____%
28) 81 is what percent of 900? _____%
29) 90 is what percent of 500? _____%
30) 82 is 20 percent of what number? _____
31) 14 is 35 percent of what number? _____
32) 90 is 6 percent of what number? _____
33) 80 is 40 percent of what number? _____
34) 90 is 15 percent of what number? _____
35) 28 is 7 percent of what number? _____
36) 54 is 18 percent of what number? _____
37) 72 is 24 percent of what number? _____

Percent Problems - Answers

Solve each problem.

1) What is 5 percent of 300? 15
2) What is 15 percent of 600? 90
3) What is 12 percent of 450? 54
4) What is 30 percent of 240? 72
5) What is 60 percent of 850? 510
6) 63 is what percent of 300? 21%
7) 80 is what percent of 400? 20%
8) 70 is what percent of 700? 10%
9) 84 is what percent of 600? 14%
10) 90 is what percent of 300? 30%
11) 24 is what percent of 150? 16%
12) 12 is what percent of 80? 15%
13) 4 is what percent of 50? 8%
14) 110 is what percent of 500? 22%
15) 16 is what percent of 400? 4%
16) 39 is what percent of 300? 13%
17) 56 is what percent of 200? 28%
18) 30 is what percent of 500? 6%
19) 84 is what percent of 700? 12%
20) 40 is what percent of 500? 8%
21) 26 is what percent of 100? 26%
22) 45 is what percent of 900? 5%
23) 60 is what percent of 400? 15%
24) 18 is what percent of 900? 2%
25) 75 is what percent of 250? 30%
26) 27 is what percent of 900? 3%
27) 49 is what percent of 700? 7%
28) 81 is what percent of 900? 9%
29) 90 is what percent of 500? 18%
30) 82 is 20 percent of what number? 410
31) 14 is 35 percent of what number? 40
32) 90 is 6 percent of what number? 1,500
33) 80 is 40 percent of what number? 200
34) 90 is 15 percent of what number? 600
35) 28 is 7 percent of what number? 400
36) 54 is 18 percent of what number? 300
37) 72 is 24 percent of what number? 300

Percent of Increase and Decrease

Solve each percent of the change word problem.

1) Bob got a raise, and his hourly wage increased from $24 to $36. What is the percent increase? _____
2) The price of gasoline rose from $2.20 to $2.42 in one month. By what percent did the gas price rise? ______
3) In a class, the number of students has been increased from 30 to 39. What is the percent increase? ______
4) The price of a pair of shoes increases from $28 to $35. What is the percent increase? _____
5) In a class, the number of students has been decreased from 24 to 18. What is the percentage decrease? _____
6) Nick got a raise, and his hourly wage increased from $50 to $55. What is the percent increase? _____
7) A coat was originally priced at $80. It went on sale for $70.40. What was the percent that the coat was discounted? ___
8) The price of a pair of shoes increases from $8 to $12. What is the percent increase? _____
9) A house was purchased in 2002 for $180,000. It is now valued at $144,000. What is the rate (percent) of depreciation for the house? _____
10) The price of gasoline rose from $3.00 to $3.15 in one month. By what percent did the gas price rise? ______

Percent of Increase and Decrease – Answers

Solve each percent of the change word problem

1) Bob got a raise, and his hourly wage increased from $24 to $36. What is the percent increase? 50%
2) The price of gasoline rose from $2.20 to $2.42 in one month. By what percent did the gas price rise? 10%
3) In a class, the number of students has been increased from 30 to 39. What is the percent increase? 30%
4) The price of a pair of shoes increases from $28 to $35. What is the percent increase? 25%
5) In a class, the number of students has been decreased from 24 to 18. What is the percentage decrease? 25%
6) Nick got a raise, and his hourly wage increased from $50 to $55. What is the percent increase? 10%
7) A coat was originally priced at $80. It went on sale for $70.40. What was the percent that the coat was discounted? 12%
8) The price of a pair of shoes increases from $8 to $12. What is the percent increase? 50%
9) A house was purchased in 2002 for $180,000. It is now valued at $144,000. What is the rate (percent) of depreciation for the house? 20%
10) The price of gasoline rose from $3.00 to $3.15 in one month. By what percent did the gas price rise? 5%

Discount, Tax and Tip

Find the missing values.

1) Original price of a computer: $400, Tax: 5%, Selling price: $_____

2) Original price of a sofa: $600, Tax: 12%, Selling price: $_____

3) Original price of a table: $550, Tax: 18%, Selling price: $_____

4) Original price of a cell phone: $700, Tax: 20%, Selling price: $_____

5) Original price of a printer: $400, Tax: 22%, Selling price: $_____

6) Original price of a computer: $600, Tax: 15%, Selling price: $_____

7) Restaurant bill: $24.00, Tip: 25%, Final amount: $_____

8) Original price of a cell phone: $300 Tax: 8%, Selling price: $_____

9) Original price of a carpet: $800, Tax: 25%, Selling price: $_____

10) Original price of a camera: $200 Discount: 35%, Selling price: $_____

11) Original price of a dress: $500 Discount: 10%, Selling price: $_____

12) Original price of a monitor: $400 Discount: 5%, Selling price: $_____

13) Original price of a laptop: $900 Discount: 20%, Selling price: $_____

14) Restaurant bill: $54.00 Tip: 20%, Final amount: $_____

Discount, Tax and Tip – Answers

Find the missing values.

1) Original price of a computer: $400 Tax: 5%, Selling price: $420
2) Original price of a sofa: $600 Tax: 12%, Selling price: $672
3) Original price of a table: $550 Tax: 18%, Selling price: $649
4) Original price of a cell phone: $700 Tax: 20%, Selling price: $840
5) Original price of a printer: $400 Tax: 22%, Selling price: $488
6) Original price of a computer: $600 Tax: 15%, Selling price: $690
7) Restaurant bill: $24.00 Tip: 25%, Final amount: $30.00
8) Original price of a cell phone: $300 Tax: 8%, Selling price: $324
9) Original price of a carpet: $800 Tax: 25%, Selling price: $1,000
10) Original price of a camera: $200 Discount: 35%, Selling price: $130
11) Original price of a dress: $500 Discount: 10%, Selling price: $450
12) Original price of a monitor: $400 Discount: 5%, Selling price: $380
13) Original price of a laptop: $900 Discount: 20%, Selling price: $720
14) Restaurant bill: $54.00 Tip: 20%, Final amount: $64.80

Simplifying Variable Expressions

Simplify and write the answer.

1) $3x + 5 + 2x =$
2) $7x + 3 - 3x =$
3) $-2 - x^2 - 6x^2 =$
4) $(-6)(8x - 4) =$
5) $3 + 10x^2 + 2x =$
6) $8x^2 + 6x + 7x^2 =$
7) $2x^2 - 5x - 7x =$
8) $x - 3 + 5 - 3x =$
9) $2 - 3x + 12 - 2x =$
10) $5x^2 - 12x^2 + 8x =$
11) $2x^2 + 6x + 3x^2 =$
12) $2x^2 - 2x - x =$
13) $2x^2 - (-8x + 6) = 2$
14) $4x + 6(2 - 5x) =$
15) $10x + 8(10x - 6) =$
16) $9(-2x - 6) - 5 =$
17) $32x - 4 + 23 + 2x =$
18) $8x - 12x - x^2 + 13 =$
19) $(-6)(8x - 4) + 10x =$
20) $14x - 5(5 - 8x) =$
21) $23x + 4(9x + 3) + 12 =$
22) $3(-7x + 5) + 20x =$
23) $12x - 3x(x + 9) =$
24) $7x + 5x(3 - 3x) =$
25) $5x(-8x + 12) + 14x =$
26) $40x + 12 + 2x^2 =$
27) $5x(x - 3) - 10 =$
28) $8x - 7 + 8x + 2x^2 =$
29) $7x - 3x^2 - 5x^2 - 3 =$
30) $4 + x^2 - 6x^2 - 12x =$
31) $12x + 8x^2 + 2x + 20 =$
32) $23 + 15x^2 + 8x - 4x^2 =$

Simplifying Variable Expressions - Answers

Simplify and write the answer.

1) $3x + 5 + 2x = 5x + 5$

2) $7x + 3 - 3x = 4x + 3$

3) $-2 - x^2 - 6x^2 = -7x^2 - 2$

4) $(-6)(8x - 4) = -48x + 24$

5) $3 + 10x^2 + 2x = 10x^2 + 2x + 3$

6) $8x^2 + 6x + 7x^2 = 15x^2 + 6x$

7) $2x^2 - 5x - 7x = 2x^2 - 12x$

8) $x - 3 + 5 - 3x = -2x + 2$

9) $2 - 3x + 12 - 2x = -5x + 14$

10) $5x^2 - 12x^2 + 8x = -7x^2 + 8x$

11) $2x^2 + 6x + 3x^2 = 5x^2 + 6x$

12) $2x^2 - 2x - x = 2x^2 - 3x$

13) $2x^2 - (-8x + 6) = 2x^2 + 8x - 6$

14) $4x + 6(2 - 5x) = -26x + 12$

15) $10x + 8(10x - 6) = 90x - 48$

16) $9(-2x - 6) - 5 = -18x - 59$

17) $32x - 4 + 23 + 2x = 34x + 19$

18) $8x - 12x - x^2 + 13 = -x^2 - 4x + 13$

19) $(-6)(8x - 4) + 10x = -38x + 24$

20) $14x - 5(5 - 8x) = 54x - 25$

21) $23x + 4(9x + 3) + 12 = 59x + 24$

22) $3(-7x + 5) + 20x = -x + 15$

23) $12x - 3x(x + 9) = -3x^2 - 15x$

24) $7x + 5x(3 - 3x) = -15x^2 + 22x$

25) $5x(-8x + 12) + 14x = -40x^2 + 74x$

26) $40x + 12 + 2x^2 = 2x^2 + 40x + 12$

27) $5x(x - 3) - 10 = 5x^2 - 15x - 10$

28) $8x - 7 + 8x + 2x^2 = 2x^2 + 16x - 7$

29) $7x - 3x^2 - 5x^2 - 3 = -8x^2 + 7x - 3$

30) $4 + x^2 - 6x^2 - 12x = -5x^2 - 12x + 4$

31) $12x + 8x^2 + 2x + 20 = 8x^2 + 14x + 20$

32) $23 + 15x^2 + 8x - 4x^2 = 11x^2 + 8x + 23$

Simplifying Polynomial Expressions

Simplify and write the answer.

1) $(2x^3 + 5x^2) - (12x + 2x^2) =$ ____________

2) $(-x^5 + 2x^3) - (3x^3 + 6x^2) =$ ____________

3) $(12x^4 + 4x^2) - (2x^2 - 6x^4) =$ ____________

4) $4x - 3x^2 - 2(6x^2 + 6x^3) =$ ____________

5) $(2x^3 - 3) + 3(2x^2 - 3x^3) =$ ____________

6) $4(4x^3 - 2x) - (3x^3 - 2x^4) =$ ____________

7) $2(4x - 3x^3) - 3(3x^3 + 4x^2) =$ ____________

8) $(2x^2 - 2x) - (2x^3 + 5x^2) =$ ____________

9) $2x^3 - (4x^4 + 2x) + x^2 =$ ____________

10) $x^4 - 9(x^2 + x) - 5x =$ ____________

11) $(-2x^2 - x^4) + (4x^4 - x^2) =$ ____________

12) $4x^2 - 5x^3 + 15x^4 - 12x^3 =$ ____________

13) $2x^2 - 5x^4 + 14x^4 - 11x^3 =$ ____________

14) $2x^2 + 5x^3 - 7x^2 + 12x =$ ____________

15) $2x^4 - 5x^5 + 8x^4 - 8x^2 =$ ____________

16) $5x^3 + 17x - 5x^2 - 2x^3 =$ ____________

Simplifying Polynomial Expressions - Answers

Simplify and write the answer.

1) $(2x^3 + 5x^2) - (12x + 2x^2) = 2x^3 + 3x^2 - 12x$

2) $(-x^5 + 2x^3) - (3x^3 + 6x^2) = -x^5 - x^3 - 6x^2$

3) $(12x^4 + 4x^2) - (2x^2 - 6x^4) = 18x^4 + 2x^2$

4) $4x - 3x^2 - 2(6x^2 + 6x^3) = -12x^3 - 15x^2 + 4x$

5) $(2x^3 - 3) + 3(2x^2 - 3x^3) = -7x^3 + 6x^2 - 3$

6) $4(4x^3 - 2x) - (3x^3 - 2x^4) = 2x^4 + 13x^3 - 8x$

7) $2(4x - 3x^3) - 3(3x^3 + 4x^2) = -15x^3 - 12x^2 + 8x$

8) $(2x^2 - 2x) - (2x^3 + 5x^2) = -2x^3 - 3x^2 - 2x$

9) $2x^3 - (4x^4 + 2x) + x^2 = -4x^4 + 2x^3 + x^2 - 2x$

10) $x^4 - 9(x^2 + x) - 5x = x^4 - 9x^2 - 14x$

11) $(-2x^2 - x^4) + (4x^4 - x^2) = 3x^4 - 3x^2$

12) $4x^2 - 5x^3 + 15x^4 - 12x^3 = 15x^4 - 17x^3 + 4x^2$

13) $2x^2 - 5x^4 + 14x^4 - 11x^3 = 9x^4 - 11x^3 + 2x^2$

14) $2x^2 + 5x^3 - 7x^2 + 12x = 5x^3 - 5x^2 + 12x$

15) $2x^4 - 5x^5 + 8x^4 - 8x^2 = -5x^5 + 10x^4 - 8x^2$

16) $5x^3 + 17x - 5x^2 - 2x^3 = 3x^3 - 5x^2 + 17x$

Evaluating One Variable

Evaluate each expression using the value given.

1) $x = 3 \Rightarrow 6x - 9 =$

2) $x = 2 \Rightarrow 7x - 10 =$

3) $x = 1 \Rightarrow 5x + 2 =$

4) $x = 2 \Rightarrow 3x + 9 =$

5) $x = 4 \Rightarrow 4x - 8 =$

6) $x = 2 \Rightarrow 5x - 2x + 10 =$

7) $x = 3 \Rightarrow 2x - x - 6 =$

8) $x = 4 \Rightarrow 6x - 3x + 4 =$

9) $x = -2 \Rightarrow 4x - 6x - 5 =$

10) $x = -1 \Rightarrow 3x - 5x + 11 =$

11) $x = 1 \Rightarrow x - 7x + 12 =$

12) $x = 2 \Rightarrow 2(-3x + 4) =$

13) $x = 3 \Rightarrow 4(-5x - 2) =$

14) $x = 2 \Rightarrow 5(-2x - 4) =$

15) $x = -2 \Rightarrow 3(-4x - 5) =$

16) $x = 3 \Rightarrow 8x + 5 =$

17) $x = -3 \Rightarrow 12x + 9 =$

18) $x = -1 \Rightarrow 9x - 8 =$

19) $x = 2 \Rightarrow 16x - 10 =$

20) $x = 1 \Rightarrow 4x + 3 =$

21) $x = 5 \Rightarrow 7x - 2 =$

22) $x = 7 \Rightarrow 28 - x =$

23) $x = 3 \Rightarrow 5x - 10 =$

24) $x = 12 \Rightarrow 40 - 2x =$

25) $x = 2 \Rightarrow 11x - 2 =$

26) $x = 3 \Rightarrow 2x - x + 10 =$

Evaluating One Variable – Answers

Evaluate each expression using the value given.

1) $x = 3 \Rightarrow 6x - 9 = 9$

2) $x = 2 \Rightarrow 7x - 10 = 4$

3) $x = 1 \Rightarrow 5x + 2 = 7$

4) $x = 2 \Rightarrow 3x + 9 = 15$

5) $x = 4 \Rightarrow 4x - 8 = 8$

6) $x = 2 \Rightarrow 5x - 2x + 10 = 16$

7) $x = 3 \Rightarrow 2x - x - 6 = -3$

8) $x = 4 \Rightarrow 6x - 3x + 4 = 16$

9) $x = -2 \Rightarrow 4x - 6x - 5 = -1$

10) $x = -1 \Rightarrow 3x - 5x + 11 = 13$

11) $x = 1 \Rightarrow x - 7x + 12 = 6$

12) $x = 2 \Rightarrow 2(-3x + 4) = -4$

13) $x = 3 \Rightarrow 4(-5x - 2) = -68$

14) $x = 2 \Rightarrow 5(-2x - 4) = -40$

15) $x = -2 \Rightarrow 3(-4x - 5) = 9$

16) $x = 3 \Rightarrow 8x + 5 = 29$

17) $x = -3 \Rightarrow 12x + 9 = -27$

18) $x = -1 \Rightarrow 9x - 8 = -17$

19) $x = 2 \Rightarrow 16x - 10 = 22$

20) $x = 1 \Rightarrow 4x + 3 = 7$

21) $x = 5 \Rightarrow 7x - 2 = 33$

22) $x = 7 \Rightarrow 28 - x = 21$

23) $x = 3 \Rightarrow 5x - 10 = 5$

24) $x = 12 \Rightarrow 40 - 2x = 16$

25) $x = 2 \Rightarrow 11x - 2 = 20$

26) $x = 3 \Rightarrow 2x - x + 10 = 13$

Evaluating Two Variables

Evaluate each expression using the values given.

1) $2x + 3y, x = 2, y = 3$

____

2) $3x + 4y, x = -1, y = -2$

____

3) $x + 6y, x = 3, y = 1$

____

4) $2a - (15 - b), a = 2, b = 3$

____

5) $4a - (6 - 3b), a = 1, b = 4$

____

6) $a - (8 - 2b), a = 2, b = 5$

____

7) $3z + 21 + 5k, z = 4, k = 1$

____

8) $-7a + 4b, a = 6, b = 3$

____

9) $-4a + 3b, a = 2, b = 4$

____

10) $-6a + 6b, a = 4, b = 3$

____

11) $-8a + 2b, a = 4, b = 6$

____

12) $4x + 6y, x = 6, y = 3$

____

13) $2x + 9y, x = 8, y = 1$

____

14) $x - 7y, x = 9, y = 4$

____

15) $5x - 4y, x = 6, y = 3$

____

16) $2z + 14 + 8k, z = 4, k = 1$

____

17) $6x + 3y, x = 3, y = 8$

____

18) $5a - 6b, a = -3, b = -1$

____

19) $8a + 4b, a = -4, b = 3$

____

20) $-2a - b, a = 4, b = 9$

____

21) $-7a + 3b, a = 4, b = 3$

____

22) $-5a + 9b, a = 7, b = 1$

____

Evaluating Two Variables - Answers

Evaluate each expression using the values given.

1) $2x + 3y, x = 2, y = 3$
   13
2) $3x + 4y, x = -1, y = -2$
   -11
3) $x + 6y, x = 3, y = 1$
   9
4) $2a - (15 - b), a = 2, b = 3$
   -8
5) $4a - (6 - 3b), a = 1, b = 4$
   10
6) $a - (8 - 2b), a = 2, b = 5$
   4
7) $3z + 21 + 5k, z = 4, k = 1$
   38
8) $-7a + 4b, a = 6, b = 3$
   -30
9) $-4a + 3b, a = 2, b = 4$
   4
10) $-6a + 6b, a = 4, b = 3$
   $- 6$
11) $-8a + 2b, a = 4, b = 6$
   -20
12) $4x + 6y, x = 6, y = 3$
   42
13) $2x + 9y, x = 8, y = 1$
   25
14) $x - 7y, x = 9, y = 4$
   -19
15) $5x - 4y, x = 6, y = 3$
   18
16) $2z + 14 + 8k, z = 4, k = 1$
   30
17) $6x + 3y, x = 3, y = 8$
   42
18) $5a - 6b, a = -3, b = -1$
   -9
19) $8a + 4b, a = -4, b = 3$
   -20
20) $-2a - b, a = 4, b = 9$
   -17
21) $-7a + 3b, a = 4, b = 3$
   -19
22) $-5a + 9b, a = 7, b = 1$
   -26

The Distributive Property

Use the distributive property to simplify each expression.

1) $(-3)(12x+3)=$

2) $(-4x+5)(-6)=$

3) $13(-4x+2)=$

4) $7(6-3x)=$

5) $(6-5x)(-4)=$

6) $9(8-2x)=$

7) $(-4x+6)5=$

8) $(-2x+7)(-8)=$

9) $8(-4x+7)=$

10) $(-9x+5)(-3)=$

11) $8(-x+9)=$

12) $7(2-6x)=$

13) $(-12x+4)(-3)=$

14) $(-6)(-10x+6)=$

15) $(-5)(5-11x)=$

16) $9(4-8x)=$

17) $(-6x+2)7=$

18) $(-9)(1-12x)=$

19) $(-3)(4-6x)=$

20) $(2-8x)(-2)=$

21) $20(2-x)=$

22) $12(-4x+3)=$

23) $15(2-3x)=$

24) $(-4x+5)2=$

25) $(-11x+8)(-2)=$

26) $14(5-8x)=$

The Distributive Property - Answers

Use the distributive property to simplify each expression.

1) $(-3)(12x+3) = -36x - 9$

2) $(-4x+5)(-6) = 24x - 30$

3) $13(-4x+2) = -52x + 26$

4) $7(6-3x) = -21x + 42$

5) $(6-5x)(-4) = 20x - 24$

6) $9(8-2x) = -18x + 72$

7) $(-4x+6)5 = -20x + 30$

8) $(-2x+7)(-8) = 16x - 56$

9) $8(-4x+7) = -32x + 56$

10) $(-9x+5)(-3) = 27x - 15$

11) $8(-x+9) = -8x + 72$

12) $7(2-6x) = -42x + 14$

13) $(-12x+4)(-3) = 36x - 12$

14) $(-6)(-10x+6) = 60x - 36$

15) $(-5)(5-11x) = 55x - 25$

16) $9(4-8x) = -72x + 36$

17) $(-6x+2)7 = -42x + 14$

18) $(-9)(1-12x) = 108x - 9$

19) $(-3)(4-6x) = 18x - 12$

20) $(2-8x)(-2) = 16x - 4$

21) $20(2-x) = -20x + 40$

22) $12(-4x+3) = -48x + 36$

23) $15(2-3x) = -45x + 30$

24) $(-4x+5)2 = -8x + 10$

25) $(-11x+8)(-2) = 22x - 16$

26) $14(5-8x) = -112x + 70$

One–Step Equations

Solve each equation for x.

1) $x - 15 = 24 \Rightarrow x =$ ____

2) $18 = -6 + x \Rightarrow x =$ ____

3) $19 - x = 8 \Rightarrow x =$ ____

4) $x - 22 = 24 \Rightarrow x =$ ____

5) $24 - x = 17 \Rightarrow x =$ ____

6) $16 - x = 3 \Rightarrow x =$ ____

7) $x + 14 = 12 \Rightarrow x =$ ____

8) $26 + x = 8 \Rightarrow x =$ ____

9) $x + 9 = -18 \Rightarrow x =$ ____

10) $x + 21 = 11 \Rightarrow x =$ ____

11) $17 = -5 + x \Rightarrow x =$ ____

12) $x + 20 = 29 \Rightarrow x =$ ____

13) $x - 13 = 19 \Rightarrow x =$ ____

14) $x + 9 = -17 \Rightarrow x =$ ____

15) $x + 4 = -23 \Rightarrow x =$ ____

16) $16 = -9 + x \Rightarrow x =$ ____

17) $4x = 28 \Rightarrow x =$ ____

18) $21 = -7x \Rightarrow x =$ ____

19) $12x = -12 \Rightarrow x =$ ____

20) $13x = 39 \Rightarrow x =$ ____

21) $8x = -16 \Rightarrow x =$ ____

22) $\frac{x}{2} = -5 \Rightarrow x =$ ____

23) $\frac{x}{9} = 6 \Rightarrow x =$ ____

24) $27 = \frac{x}{5} \Rightarrow x =$ ____

25) $\frac{x}{4} = -3 \Rightarrow x =$ ____

26) $x \div 8 = 7 \Rightarrow x =$ ____

27) $x \div 2 = -3 \Rightarrow x =$ ____

28) $4x = 48 \Rightarrow x =$ ____

29) $9x = 72 \Rightarrow x =$ ____

30) $8x = -32 \Rightarrow x =$ ____

31) $80 = -10x \Rightarrow x =$ ____

One – Step Equations - Answers

Solve each equation for x.

1) $x - 15 = 24 \Rightarrow x = 39$

2) $18 = -6 + x \Rightarrow x = 24$

3) $19 - x = 8 \Rightarrow x = 11$

4) $x - 22 = 24 \Rightarrow x = 46$

5) $24 - x = 17 \Rightarrow x = 7$

6) $16 - x = 3 \Rightarrow x = 13$

7) $x + 14 = 12 \Rightarrow x = 26$

8) $26 + x = 8 \Rightarrow x = -18$

9) $x + 9 = -18 \Rightarrow x = -27$

10) $x + 21 = 11 \Rightarrow x = -10$

11) $17 = -5 + x \Rightarrow x = 22$

12) $x + 20 = 29 \Rightarrow x = 9$

13) $x - 13 = 19 \Rightarrow x = 32$

14) $x + 9 = -17 \Rightarrow x = -26$

15) $x + 4 = -23 \Rightarrow x = -27$

16) $16 = -9 + x \Rightarrow x = 25$

17) $4x = 28 \Rightarrow x = 7$

18) $21 = -7x \Rightarrow x = -3$

19) $12x = -12 \Rightarrow x = -1$

20) $13x = 39 \Rightarrow x = 3$

21) $8x = -16 \Rightarrow x = -2$

22) $\frac{x}{2} = -5 \Rightarrow x = -10$

23) $\frac{x}{9} = 6 \Rightarrow x = 54$

24) $27 = \frac{x}{5} \Rightarrow x = 135$

25) $\frac{x}{4} = -3 \Rightarrow x = -12$

26) $x \div 8 = 7 \Rightarrow x = 56$

27) $x \div 2 = -3 \Rightarrow x = -6$

28) $4x = 48 \Rightarrow x = 12$

29) $9x = 72 \Rightarrow x = 8$

30) $8x = -32 \Rightarrow x = -4$

31) $80 = -10x \Rightarrow x = -8$

Multi – Step Equations

Solve each equation.

1) $3x - 8 = 13 \Rightarrow x =$ ____

2) $23 = -(x - 5) \Rightarrow x =$ ____

3) $-(8 - x) = 15 \Rightarrow x =$ ____

4) $29 = -x + 12 \Rightarrow x =$ ____

5) $2(3 - 2x) = 10 \Rightarrow x =$ ____

6) $3x - 3 = 15 \Rightarrow x =$ ____

7) $32 = -x + 15 \Rightarrow x =$ ____

8) $-(10 - x) = -13 \Rightarrow x =$ ____

9) $-4(7 + x) = 4 \Rightarrow x =$ ____

10) $23 = 2x - 8 \Rightarrow x =$ ____

11) $-6(3 + x) = 6 \Rightarrow x =$ ____

12) $-3 = 3x - 15 \Rightarrow x =$ ____

13) $-7(12 + x) = 7 \Rightarrow x =$ ____

14) $8(6 - 4x) = 16 \Rightarrow x =$ ____

15) $18 - 4x = -9 - x \Rightarrow x =$ ____

16) $6(4 - x) = 30 \Rightarrow x =$ ____

17) $15 - 3x = -5 - x \Rightarrow x =$ ____

18) $9(-7 - 3x) = 18 \Rightarrow x =$ ____

19) $16 - 2x = -4 - 7x \Rightarrow x =$ ____

20) $14 - 2x = 14 + x \Rightarrow x =$ ____

21) $21 - 3x = -7 - 10x \Rightarrow x =$ ___

22) $8 - 2x = 11 + x \Rightarrow x =$ ____

23) $10 + 12x = -8 + 6x \Rightarrow x =$ ____

24) $25 + 20x = -5 + 5x \Rightarrow x =$ ____

25) $16 - x = -8 - 7x \Rightarrow x =$ ____

26) $17 - 3x = 13 + x \Rightarrow x =$ ____

27) $22 + 5x = -8 - x \Rightarrow x =$ ____

28) $-9(7 + x) = 9 \Rightarrow x =$ ____

29) $11 + 3x = -4 - 2x \Rightarrow x =$ ____

30) $13 - 2x = 3 - 3x \Rightarrow x =$ ____

31) $19 - x = -1 - 11x \Rightarrow x =$ ____

32) $12 - 2x = -2 - 4x \Rightarrow x =$ ____

Multi –Step Equations - Answers

Solve each equation.

1) $3x - 8 = 13 \Rightarrow x = 7$
2) $23 = -(x - 5) \Rightarrow x = -18$
3) $-(8 - x) = 15 \Rightarrow x = 23$
4) $29 = -x + 12 \Rightarrow x = -17$
5) $2(3 - 2x) = 10 \Rightarrow x = -1$
6) $3x - 3 = 15 \Rightarrow x = 6$
7) $32 = -x + 15 \Rightarrow x = -17$
8) $-(10 - x) = -13 \Rightarrow x = -3$
9) $-4(7 + x) = 4 \Rightarrow x = -8$
10) $23 = 2x - 8 \Rightarrow x = 15.5$
11) $-6(3 + x) = 6 \Rightarrow x = -4$
12) $-3 = 3x - 15 \Rightarrow x = 4$
13) $-7(12 + x) = 7 \Rightarrow x = -13$
14) $8(6 - 4x) = 16 \Rightarrow x = 1$
15) $18 - 4x = -9 - x \Rightarrow x = 9$
16) $6(4 - x) = 30 \Rightarrow x = -1$
17) $15 - 3x = -5 - x \Rightarrow x = 10$
18) $9(-7 - 3x) = 18 \Rightarrow x = -3$
19) $16 - 2x = -4 - 7x \Rightarrow x = -4$
20) $14 - 2x = 14 + x \Rightarrow x = 0$
21) $21 - 3x = -7 - 10x \Rightarrow x = -4$
22) $8 - 2x = 11 + x \Rightarrow x = -1$
23) $10 + 12x = -8 + 6x \Rightarrow x = -3$
24) $25 + 20x = -5 + 5x \Rightarrow x = -2$
25) $16 - x = -8 - 7x \Rightarrow x = -4$
26) $17 - 3x = 13 + x \Rightarrow x = 1$
27) $22 + 5x = -8 - x \Rightarrow x = -5$
28) $-9(7 + x) = 9 \Rightarrow x = -8$
29) $11 + 3x = -4 - 2x \Rightarrow x = -3$
30) $13 - 2x = 3 - 3x \Rightarrow x = -10$
31) $19 - x = -1 - 11x \Rightarrow x = -2$
32) $12 - 2x = -2 - 4x \Rightarrow x = -7$

Graphing Single–Variable Inequalities

Graph each inequality

1) $\boldsymbol{x < 6}$

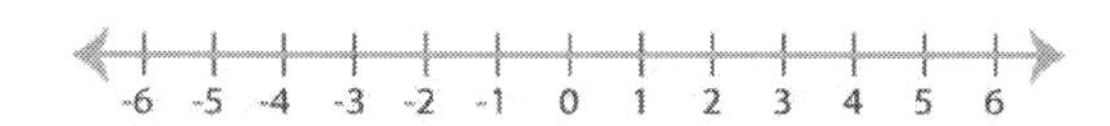

2) $\boldsymbol{x \geq 1}$

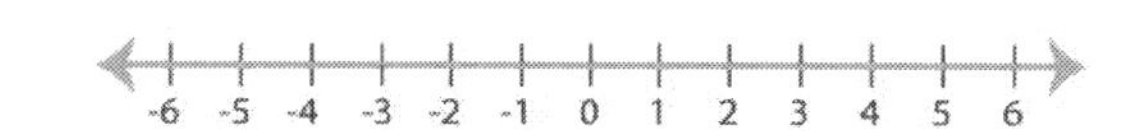

3) $\boldsymbol{x \geq -6}$

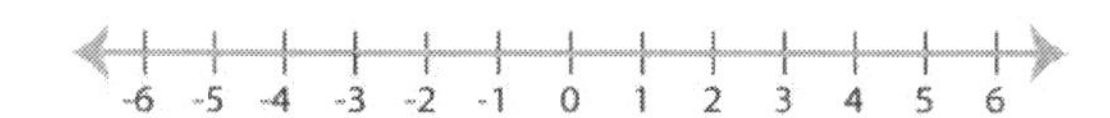

4) $x \leq -2$

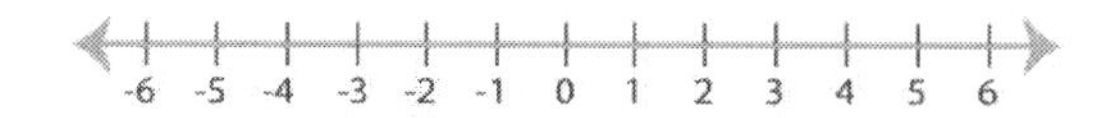

5) $\mathbf{x > -1}$

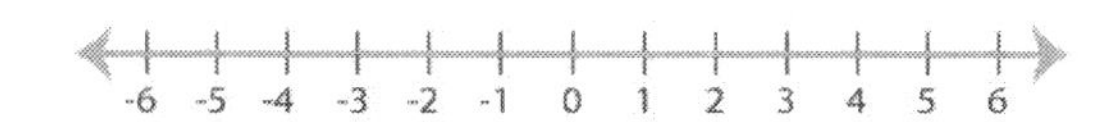

6) $3 > x$

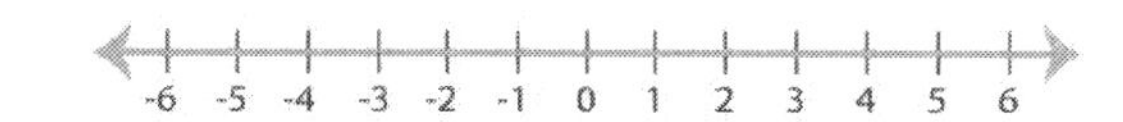

7) $\mathbf{2 \leq x}$

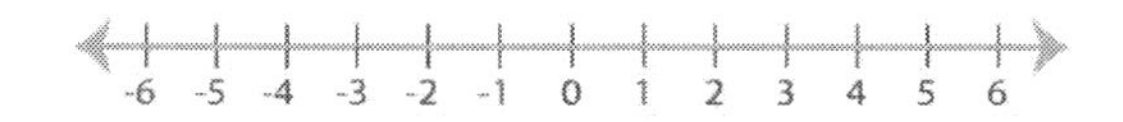

8) $x > 0$

9) $\mathbf{-3 \leq x}$

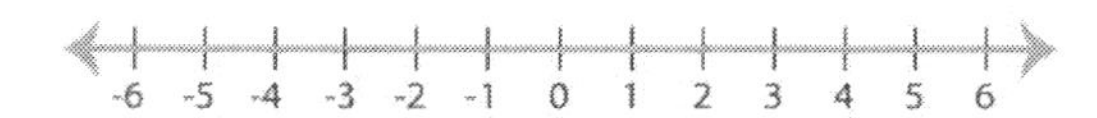

10) $-4 \leq x$

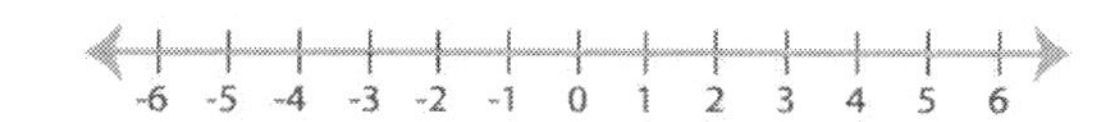

11) $\mathbf{x \leq 5}$

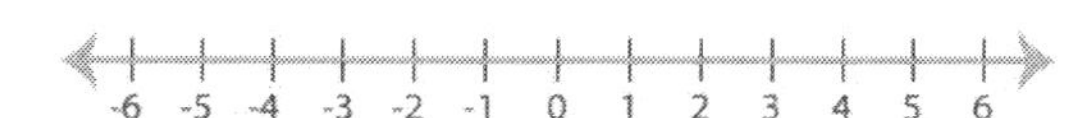

12) $0 \leq x$

13) $\mathbf{-5 \leq x}$

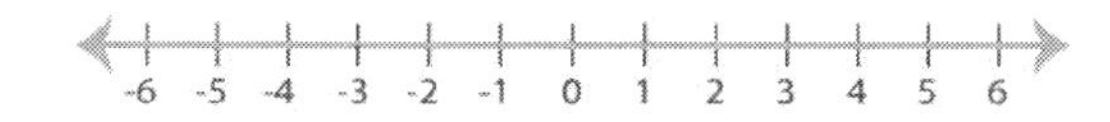

14) $x > -6$

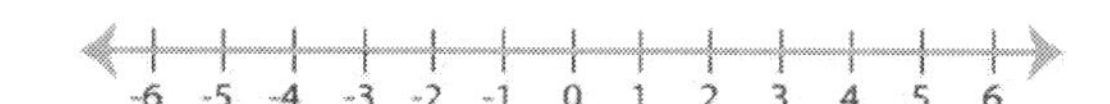

Graphing Single–Variable Inequalities - Answers

Graph each inequality.

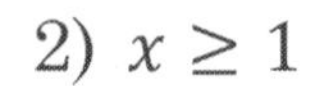

1) $x < 6$

2) $x \geq 1$

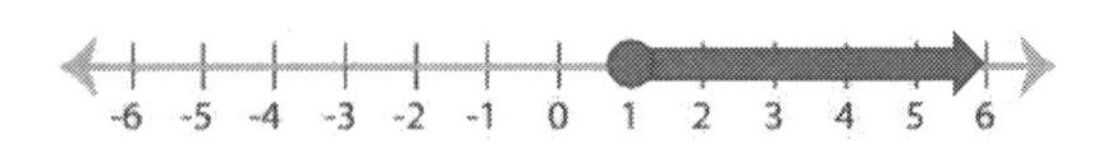

3) $x \geq -6$

4) $x \leq -2$

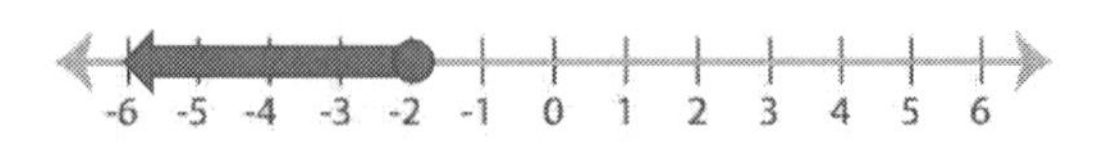

5) $x > -1$

6) $3 > x$

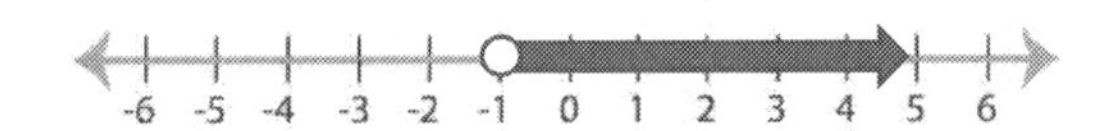

7) $2 \leq x$

8) $x > 0$

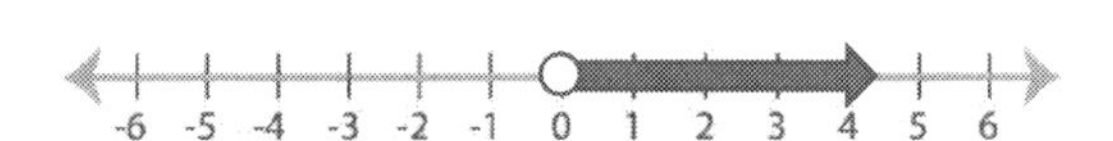

9) $-3 \leq x$

10) $-4 \leq x$

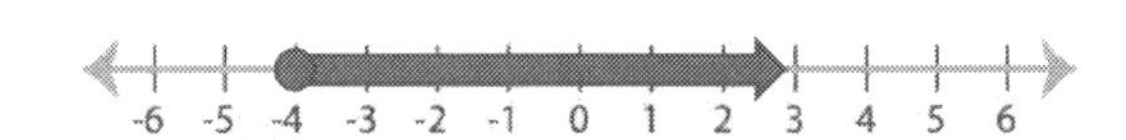

11) $x \leq 5$

12) $0 \leq \quad x$

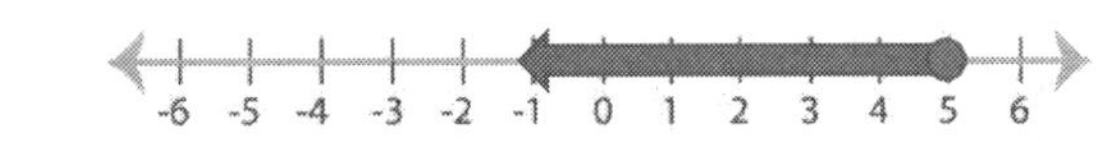

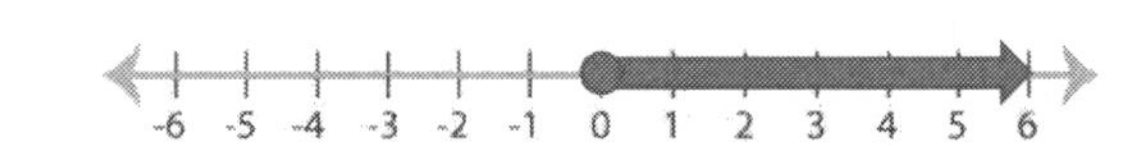

13) $-5 \leq x$

14) $x > -6$

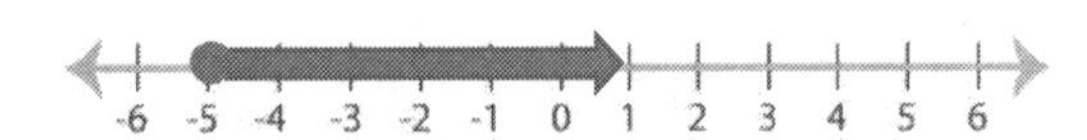

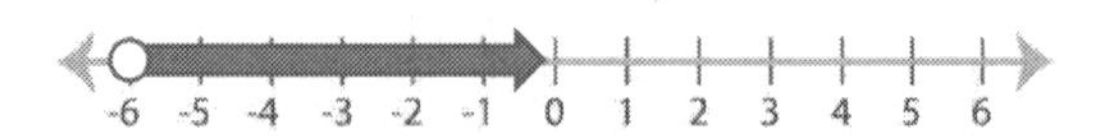

One–Step Inequalities

Solve each inequality for x.

1) $x - 10 < 22 \Rightarrow$ ______

2) $18 \leq -4 + x \Rightarrow$ ______

3) $x - 33 > 8 \Rightarrow$ ______

4) $x + 22 \geq 24 \Rightarrow$ ______

5) $x - 24 > 17 \Rightarrow$ ______

6) $x + 5 \geq 3 \Rightarrow x$______

7) $x + 14 < 12 \Rightarrow$ ______

8) $26 + x \leq 8 \Rightarrow$ ______

9) $x + 9 \geq -18 \Rightarrow$ ______

10) $x + 24 < 11 \Rightarrow$ ______

11) $17 \leq -5 + x \Rightarrow$ ______

12) $x + 25 > 29 \Rightarrow x$______

13) $x - 17 \geq 19 \Rightarrow$ ______

14) $x + 8 > -17 \Rightarrow$ ______

15) $x + 8 < -23 \Rightarrow$ ______

16) $16 \leq -5 + x \Rightarrow$ ______

17) $4x \leq 12 \Rightarrow$ ______

18) $28 \geq -7x \Rightarrow$ ______

19) $2x > -14 \Rightarrow$ ______

20) $13x \leq 39 \Rightarrow$ ______

21) $-8x > -16 \Rightarrow$ ______

22) $\frac{x}{2} < -6 \Rightarrow$ ______

23) $\frac{x}{6} > 6 \Rightarrow$ ______

24) $27 \leq \frac{x}{4} \Rightarrow$ ______

25) $\frac{x}{8} < -3 \Rightarrow$ ______

26) $6x \geq 18 \Rightarrow$ ______

27) $5x \geq -25 \Rightarrow$ ______

28) $4x > 48 \Rightarrow$ ______

29) $8x \leq 72 \Rightarrow$ ______

30) $-4x < -32 \Rightarrow$ ______

31) $40 > -10x \Rightarrow$ ______

One–Step Inequalities - Answers

Solve each inequality for x.

1) $x - 10 < 22 \Rightarrow x < 32$

2) $18 \leq -4 + x \Rightarrow 22 \leq x$

3) $x - 33 > 8 \Rightarrow x > 41$

4) $x + 22 \geq 24 \Rightarrow x \geq 2$

5) $x - 24 > 17 \Rightarrow x > 41$

6) $x + 5 \geq 3 \Rightarrow x \geq -2$

7) $x + 14 < 12 \Rightarrow x < -2$

8) $26 + x \leq 8 \Rightarrow x \leq -18$

9) $x + 9 \geq -18 \Rightarrow x \geq -27$

10) $x + 24 < 11 \Rightarrow x < -13$

11) $17 \leq -5 + x \Rightarrow 22 \leq x$

12) $x + 25 > 29 \Rightarrow x > 4$

13) $x - 17 \geq 19 \Rightarrow x \geq 36$

14) $x + 8 > -17 \Rightarrow x > -25$

15) $x + 8 < -23 \Rightarrow x < -31$

16) $16 \leq -5 + x \Rightarrow 21 \leq x$

17) $4x \leq 12 \Rightarrow x \leq 3$

18) $28 \geq -7x \Rightarrow -4 \leq x$

19) $2x > -14 \Rightarrow x > -7$

20) $13x \leq 39 \Rightarrow x \leq 3$

21) $-8x > -16 \Rightarrow x < 2$

22) $\frac{x}{2} < -6 \Rightarrow x < -12$

23) $\frac{x}{6} > 6 \Rightarrow x > 36$

24) $27 \leq \frac{x}{4} \Rightarrow 108 \leq x$

25) $\frac{x}{8} < -3 \Rightarrow x < -24$

26) $6x \geq 18 \Rightarrow x \geq 3$

27) $5x \geq -25 \Rightarrow x \geq -5$

28) $4x > 48 \Rightarrow x > 12$

29) $8x \leq 72 \Rightarrow x \leq 9$

30) $-4x < -32 \Rightarrow x > 8$

31) $40 > -10x \Rightarrow -4 < x$

Multi –Step Inequalities

Solve each inequality.

1) $2x - 8 \leq 8 \rightarrow$ ______

2) $3 + 2x \geq 17 \rightarrow$ ______

3) $5 + 3x \geq 26 \rightarrow$ ______

4) $2x - 8 \leq 14 \rightarrow$ ______

5) $3x - 4 \leq 23 \rightarrow$ ______

6) $7x - 5 \leq 51 \rightarrow$ ______

7) $4x - 9 \leq 27 \rightarrow$ ______

8) $6x - 11 \leq 13 \rightarrow$ ______

9) $5x - 7 \leq 33 \rightarrow$ ______

10) $6 + 2x \geq 28 \rightarrow$ ______

11) $8 + 3x \geq 35 \rightarrow$ ______

12) $4 + 6x < 34 \rightarrow$ ______

13) $3 + 2x \geq 53 \rightarrow$ ______

14) $7 - 6x > 56 + x \rightarrow$ ______

15) $9 + 4x \geq 39 + 2x \rightarrow$ ______

16) $3 + 5x \geq 43 \rightarrow$ ______

17) $4 - 7x < 60 \rightarrow$ ______

18) $11 - 4x \geq 55 \rightarrow$ ______

19) $12 + x \geq 48 - 2x \rightarrow$ ______

20) $10 - 10x \leq -20 \rightarrow$ ______

21) $5 - 9x \geq -40 \rightarrow$ ______

22) $8 - 7x \geq 36 \rightarrow$ ______

23) $5 + 11x < 69 + 3x \rightarrow$ ______

24) $6 + 8x < 28 - 3x \rightarrow$ ______

25) $9 + 11x < 57 - x \rightarrow$ ______

26) $3 + 10x \geq 45 - 4x \rightarrow$ ______

Multi –Step Inequalities - Answers

Solve each inequality.

1) $2x - 8 \leq 8 \rightarrow x \leq 8$
2) $3 + 2x \geq 17 \rightarrow x \geq 7$
3) $5 + 3x \geq 26 \rightarrow x \geq 7$
4) $2x - 8 \leq 14 \rightarrow x \leq 11$
5) $3x - 4 \leq 23 \rightarrow x \leq 9$
6) $7x - 5 \leq 51 \rightarrow x \leq 8$
7) $4x - 9 \leq 27 \rightarrow x \leq 9$
8) $6x - 11 \leq 13 \rightarrow x \leq 4$
9) $5x - 7 \leq 33 \rightarrow x \leq 8$
10) $6 + 2x \geq 28 \rightarrow x \geq 11$
11) $8 + 3x \geq 35 \rightarrow x \geq 9$
12) $4 + 6x < 34 \rightarrow x < 5$
13) $3 + 2x \geq 53 \rightarrow x \geq 25$
14) $7 - 6x > 56 + x \rightarrow x < -7$
15) $9 + 4x \geq 39 + 2x \rightarrow x \geq 15$
16) $3 + 5x \geq 43 \rightarrow x \geq 8$
17) $4 - 7x < 60 \rightarrow x > -8$
18) $11 - 4x \geq 55 \rightarrow x \leq -11$
19) $12 + x \geq 48 - 2x \rightarrow x \geq 12$
20) $10 - 10x \leq -20 \rightarrow x \geq 3$
21) $5 - 9x \geq -40 \rightarrow x \leq 5$
22) $8 - 7x \geq 36 \rightarrow x \leq -4$
23) $5 + 11x < 69 + 3x \rightarrow x < 8$
24) $6 + 8x < 28 - 3x \rightarrow x < 2$
25) $9 + 11x < 57 - x \rightarrow x < 4$
26) $3 + 10x \geq 45 - 4x \rightarrow x \geq 3$

System of Equations

Solve each system of equations.

1) $-x + y = 2$ $\quad x =$
$-2x + y = 3$ $\quad y =$

2) $-5x + y = -3$ $\quad x =$
$3x - 8y = 24$ $\quad y =$

3) $y = -5$ $\quad x =$
$4x - 5y = 13$

4) $y = -6x + 8$ $\quad x =$
$5x - 4y = -3$ $\quad y =$

5) $10x - 8y = -15$ $\quad x =$
$-6x + 4y = 13$ $\quad y =$

6) $-3x - 4y = 5$ $\quad x =$
$x - 2y = 5$ $\quad y =$

7) $5x - 12y = -19$ $\quad x =$
$-6x + 7y = 8$ $\quad y =$

8) $5x - 7y = -2$ $\quad x =$
$-x - 2y = -3$ $\quad y =$

9) $-x + 3y = 3$ $\quad x =$
$-7x + 8y = -5$ $\quad y =$

10) $-4x + 3y = -18$ $\quad x =$
$4x - y = 14$ $\quad y =$

11) $6x - 7y = -8$ $\quad x =$
$-x - 4y = -9$ $\quad y =$

12) $-3x + 2y = -16$ $\quad x =$
$4x - y = 13$ $\quad y =$

System of Equations- Answers

Solve each system of equations.

1) $-x + y = 2$, $-2x + y = 3$ — $x = -1$, $y = 1$

2) $-5x + y = -3$, $3x - 8y = 24$ — $x = 0$, $y = -3$

3) $y = -5$, $4x - 5y = 13$ — $x = -3$

4) $y = -6x + 8$, $5x - 4y = -3$ — $x = 1$, $y = 2$

5) $10x - 8y = -15$, $-6x + 4y = 13$ — $x = -\frac{11}{2}$, $y = -5$

6) $-3x - 4y = 5$, $x - 2y = 5$ — $x = 1$, $y = -2$

7) $5x - 12y = -19$, $-6x + 7y = 8$ — $x = 1$, $y = 2$

8) $5x - 7y = -2$, $-x - 3y = -3$ — $x = 1$, $y = 1$

9) $-x + 3y = 3$, $-7x + 8y = -5$ — $x = 3$, $y = 2$

10) $-4x + 3y = -18$, $4x - y = 14$ — $x = 3$, $y = -2$

11) $6x - 7y = -8$, $-x - 4y = -9$ — $x = 1$, $y = 2$

12) $-3x + 2y = -16$, $4x - y = 13$ — $x = 2$, $y = -5$

Finding Slope

Find the slope of each line.

1) $y = x - 5$, Slope $=$
2) $y = -3x + 2$, Slope $=$
3) $y = -x - 1$, Slope $=$
4) $y = -x - 9$, Slope $=$
5) $y = 5 + 2x$, Slope $=$
6) $y = 1 - 8x$, Slope $=$
7) $y = -4x + 3$, Slope $=$
8) $y = -9x + 8$, Slope $=$
9) $y = -2x + 4$, Slope $=$
10) $y = 9x - 8$, Slope $=$
11) $y = \frac{1}{2}x + 4$, Slope $=$
12) $y = -\frac{2}{5}x + 7$, Slope $=$
13) $-x + 3y = 5$, Slope $=$
14) $4x + 4y = 6$, Slope $=$
15) $6y - 2x = 10$, Slope $=$
16) $7y - x = 2$, Slope $=$

Find the slope of the line through each pair of points.

1) $(4,4), (8,12)$, Slope $=$
2) $(-2,4), (0,6)$, Slope $=$
3) $(6,-2), (2,6)$, Slope $=$
4) $(-4,-2), (0,6)$, Slope $=$
5) $(6,2), (3,5)$, Slope $=$
6) $(-5,1), (-1,9)$, Slope $=$
7) $(8,4), (9,6)$, Slope $=$
8) $(10,-1), (7,8)$, Slope $=$
9) $(14,-7), (13,-6)$, Slope $=$
10) $(10,7), (8,1)$, Slope $=$
11) $(5,1), (8,10)$, Slope $=$
12) $(9,-10), (8,12)$, Slope $=$

Finding Slope - answer

Find the slope of each line.

1) $y = x - 5$, Slope $= 1$
2) $y = -3x + 2$, Slope $= -3$
3) $y = -x - 1$, Slope $= -1$
4) $y = -x - 9$, Slope $= -1$
5) $y = 5 + 2x$, Slope $= 2$
6) $y = 1 - 8x$, Slope $= -8$
7) $y = -4x + 3$, Slope $= -4$
8) $y = -9x + 8$, Slope $= -9$
9) $y = -2x + 4$, Slope $= -2$
10) $y = 9x - 8$, Slope $= 9$
11) $y = \frac{1}{2}x + 4$, Slope $= 0.5$
12) $y = -\frac{2}{5}x + 7$, Slope $= -0.4$
13) $-x + 3y = 5$, Slope $= \frac{1}{3}$
14) $4x + 4y = 6$, Slope $= -1$
15) $6y - 2x = 10$, Slope $= \frac{1}{3}$
16) $7y - x = 2$, Slope $= \frac{1}{7}$

Find the slope of the line through each pair of points.

1) $(4, 4), (8, 12)$, Slope $= 2$
2) $(-2, 4), (0, 6)$, Slope $= 1$
3) $(6, -2), (2, 6)$, Slope $= -2$
4) $(-4, -2), (0, 6)$, Slope $= 2$
5) $(6, 2), (3, 5)$, Slope $= -1$
6) $(-5,1), (-1, 9)$, Slope $= 2$
7) $(8, 4), (9, 6)$, Slope $= 2$
8) $(10, -1), (7, 8)$, Slope $= -3$
9) $(14, -7), (13, -6)$, Slope $= -1$
10) $(10,7), (8, 1)$, Slope $= 3$
11) $(5, 1), (8, 10)$, Slope $= 3$
12) $(9, -10), (8, 12)$, Slope $= -22$

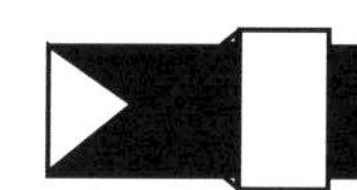

Graphing Lines Using Slope–Intercept Form

Sketch the graph of each line.

1) $\mathbf{y = -x + 1}$

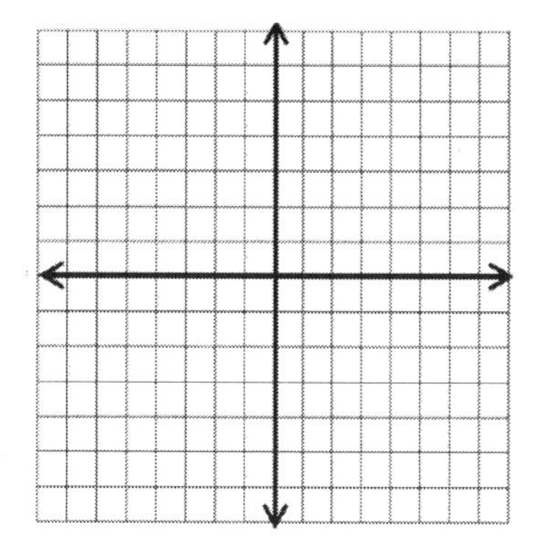

2) $\mathbf{y = 2x - 3}$

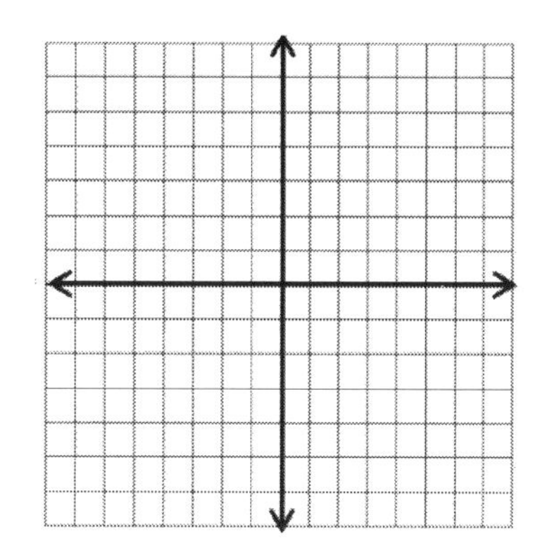

3) $\mathbf{y = -x + 2}$

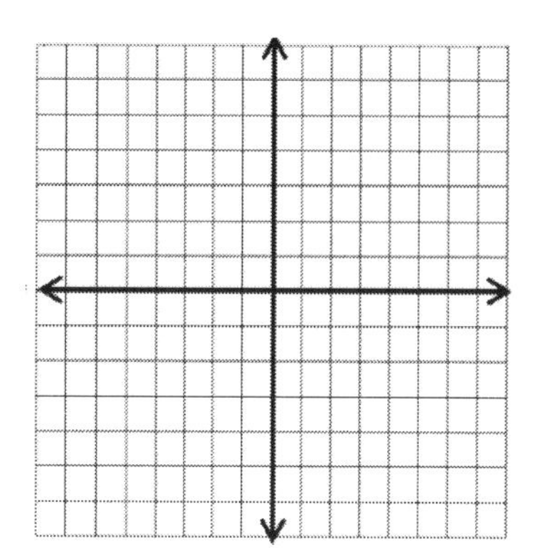

4) $\mathbf{y = x + 1}$

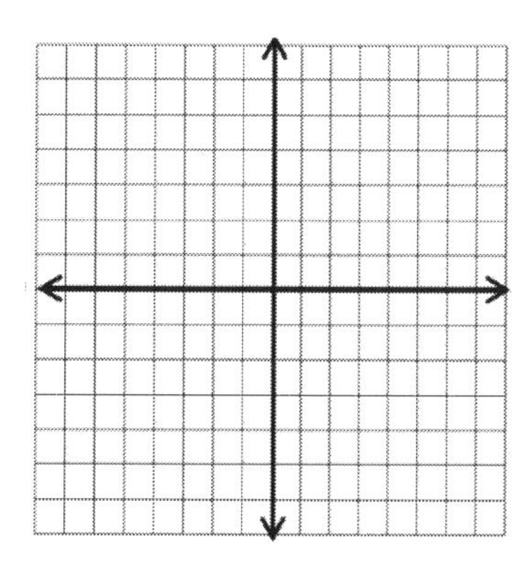

5) $y = 2x - 4$

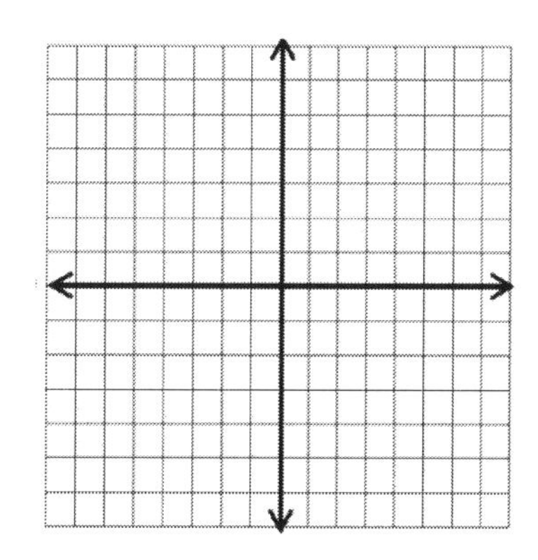

6) $y = -\frac{1}{2}x + 1$

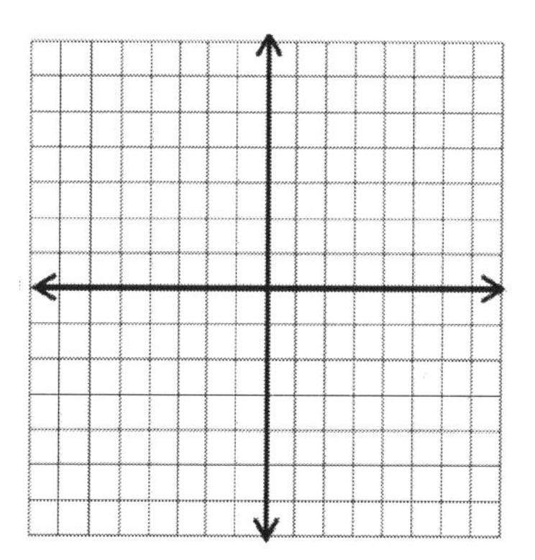

Graphing Lines Using Slope–Intercept Form - Answers

Sketch the graph of each line.

1) $y = -x + 1$

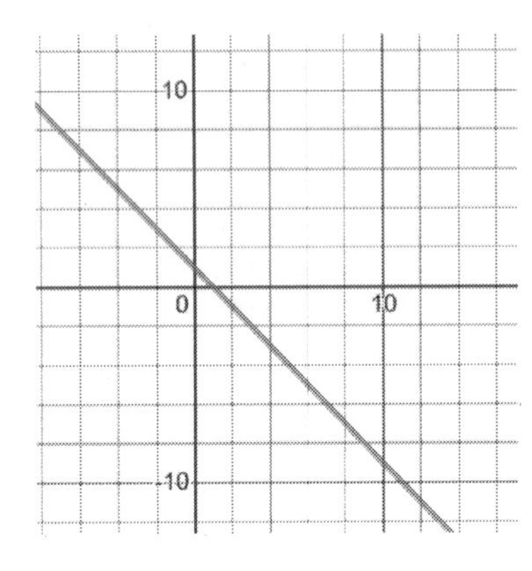

2) $y = 2x - 3$

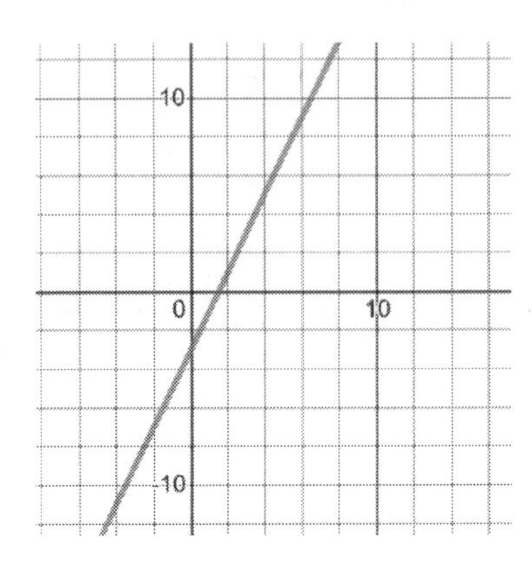

3) $y = -x + 2$

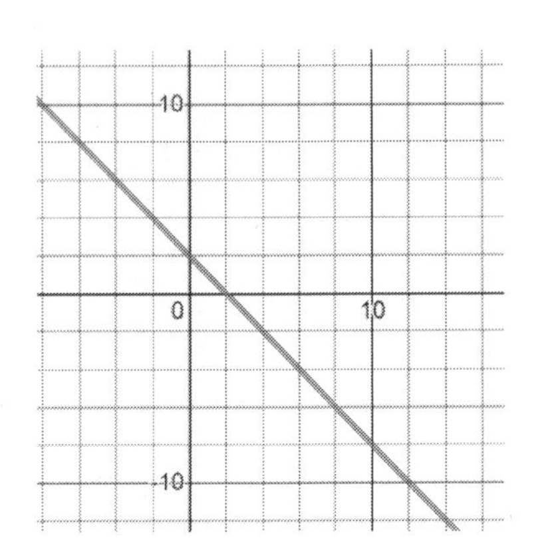

4) $y = x + 1$

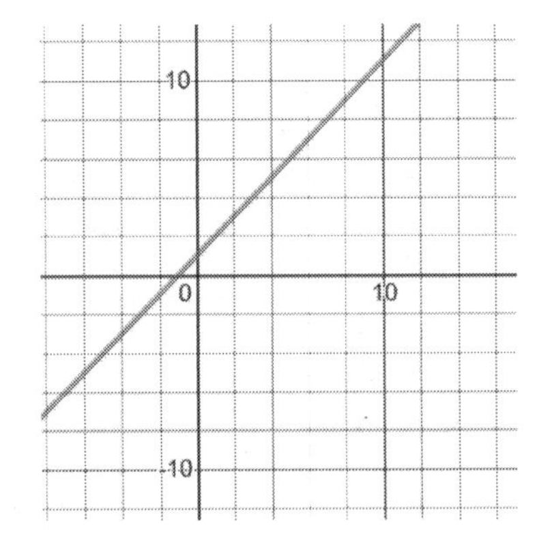

5) $y = 2x - 4$

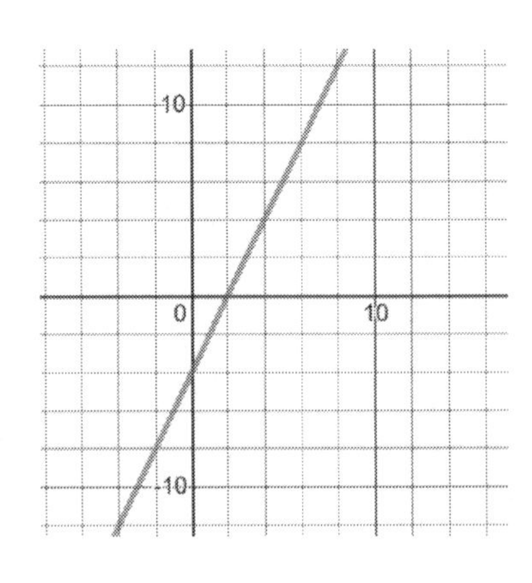

6) $y = -\frac{1}{2}x + 1$

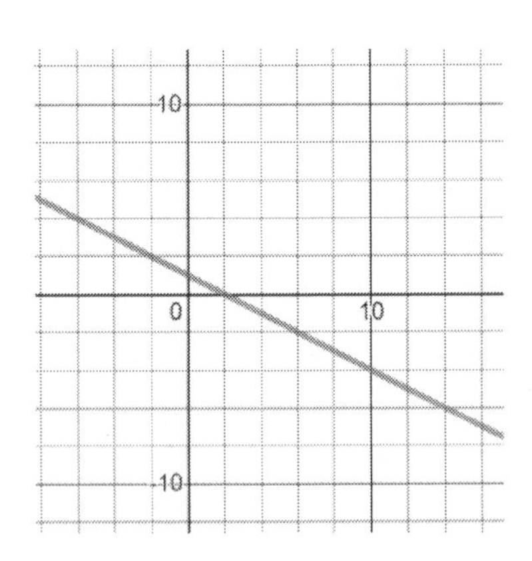

Writing Linear Equations

Write the equation of the line through the given points.

1) through: $(1,-2),(2,4)$

$y=$

2) through: $(-2,3),(1,6)$

$y=$

3) through: $(-1,2),(3,6)$

$y=$

4) through: $(8,5),(5,2)$

$y=$

5) through: $(7,-10),(2,10)$

$y=$

6) through: $(7,2),(6,1)$

$y=$

7) through: $(6,-1),(4,1)$

$y=$

8) through: $(-2,8),(-4,-6)$

$y=$

9) through: $(-2,5),(-3,4)$

$y=$

10) through: $(6,8),(8,-6)$

$y=$

11) through: $(-2,5),(-4,-3)$

$y=$

12) through: $(8,8),(4,-8)$

$y=$

13) through: $(7,-4)$, Slope: -1

$y=$

14) through: $(4,-10)$, Slope: -2

$y=$

15) through: $(6,10)$, Slope: 9

$y=$

16) through: $(-6,8)$, Slope: -2

$y=$

Solve each problem.

1) What is the equation of a line with slope 8 and intercept 5? _________
2) What is the equation of a line with slope 4 and intercept 10? _________
3) What is the equation of a line with slope 9 and passes through point $(5,23)$? _________
4) What is the equation of a line with slope -7 and passes through point $(-3,18)$? _________

Writing Linear Equations - Answers

Write the equation of the line through the given points.

1) through: $(1,-2),(2,4)$
   $y = 6x - 8$
2) through: $(-2,3),(1,6)$
   $y = x + 5$
3) through: $(-1,2),(3,6)$
   $y = x + 3$
4) through: $(8,5),(5,2)$
   $y = x - 3$
5) through: $(7,-10),(2,10)$
   $y = -4x + 18$
6) through: $(7,2),(6,1)$
   $y = x - 5$
7) through: $(6,-1),(4,1)$
   $y = -x + 5$
8) through: $(-2,8),(-4,-6)$
   $y = 7x + 22$
9) through: $(-2,5),(-3,4)$
   $y = x + 7$
10) through: $(6,8),(8,-6)$
   $y = -7x + 50$
11) through: $(-2,5),(-4,-3)$
   $y = 4x + 13$
12) through: $(8,8),(4,-8)$
   $y = 4x - 24$
13) through: $(7,-4)$, Slope: -1
   $y = -x + 3$
14) through: $(4,-10)$, Slope: -2
   $y = -2x - 2$
15) through: $(6,10)$, Slope: 9
   $y = 9x - 44$
16) through: $(-6,8)$, Slope: -2
   $y = -2x - 4$

Solve each problem.

1) What is the equation of a line with slope 8 and intercept 5? $y = 8x + 5$
2) What is the equation of a line with slope 4 and intercept 10? $y = 4x + 10$
3) What is the equation of a line with slope 9 and passes through point $(5,23)$? $y = 9x - 22$
4) What is the equation of a line with slope -7 and passes through point $(-3,18)$? $y = -7x - 3$

Finding Midpoint

Find the midpoint of the line segment with the given endpoints.

1) $(2,2),(0,4)$,

$midpoint = (\_\_,\_\_)$

2) $(3,3),(-1,5)$,

$midpoint = (\_\_,\_\_)$

3) $(2,-1),(0,5)$,

$midpoint = (\_\_,\_\_)$

4) $(-3,7),(-1,5)$,

$midpoint = (\_\_,\_\_)$

5) $(5,-2),(9,-6)$,

$midpoint = (\_\_,\_\_)$

6) $(-6,-3),(4,-7)$,

$midpoint = (\_\_,\_\_)$

7) $(7,0),(-7,8)$,

$midpoint = (\_\_,\_\_)$

8) $(-8,4),(-4,0)$,

$midpoint = (\_\_,\_\_)$

9) $(-3,6),(9,-8)$,

$midpoint = (\_\_,\_\_)$

10) $(6,8),(6,-6)$,

$midpoint = (\_\_,\_\_)$

11) $(6,7),(-8,5)$,

$midpoint = (\_\_,\_\_)$

12) $(9,3),(-3,-9)$,

$midpoint = (\_\_,\_\_)$

13) $(-6,12),(-4,6)$,

$midpoint = (\_\_,\_\_)$

14) $(10,7),(8,-3)$,

$midpoint = (\_\_,\_\_)$

15) $(13,7),(-5,3)$,

$midpoint = (\_\_,\_\_)$

16) $(-9,-4),(-5,8)$,

$midpoint = (\_\_,\_\_)$

17) $(11,7),(5,13)$,

$midpoint = (\_\_,\_\_)$

18) $(-7,-10),(11,-2)$,

$midpoint = (\_\_,\_\_)$

19) $(10,15),(-4,9)$,

$midpoint = (\_\_,\_\_)$

20) $(11,-4),(7,12)$,

$midpoint = (\_\_,\_\_)$

Finding Midpoint - Answers

Find the midpoint of the line segment with the given endpoints.

1) $(2,2),(0,4)$,
   $midpoint = (1,3)$
2) $(3,3),(-1,5)$,
   $midpoint = (1,4)$
3) $(2,-1),(0,5)$,
   $midpoint = (1,2)$
4) $(-3,7),(-1,5)$,
   $midpoint = (-2,6)$
5) $(5,-2),(9,-6)$,
   $midpoint = (7,-4)$
6) $(-6,-3),(4,-7)$,
   $midpoint = (-1,-5)$
7) $(7,0),(-7,8)$,
   $midpoint = (0,4)$
8) $(-8,4),(-4,0)$,
   $midpoint = (-6,2)$
9) $(-3,6),(9,-8)$,
   $midpoint = (3,-1)$
10) $(6,8),(6,-6)$,
   $midpoint = (6,1)$
11) $(6,7),(-8,5)$,
   $midpoint = (-1,6)$
12) $(9,3),(-3,-9)$,
   $midpoint = (3,-3)$
13) $(-6,12),(-4,6)$,
   $midpoint = (-5,9)$
14) $(10,7),(8,-3)$,
   $midpoint = (9,2)$
15) $(13,7),(-5,3)$,
   $midpoint = (4,5)$
16) $(-9,-4),(-5,8)$,
   $midpoint = (-7,2)$
17) $(11,7),(5,13)$,
   $midpoint = (8,10)$
18) $(-7,-10),(11,-2)$,
   $midpoint = (2,-6)$
19) $(10,15),(-4,9)$,
   $midpoint = (3,12)$
20) $(11,-4),(7,12)$,
   $midpoint = (9,4)$

Finding Distance of Two Points

Find the distance of each pair of points.

1) $(1, 9), (5, 6)$,

Distance = ____

2) $(-4, 5), (8, 10)$,

Distance = ____

3) $(5, -2), (-3, 4)$,

Distance = ____

4) $(-3, 0), (3, 8)$,

Distance = ____

5) $(-5, 3), (4, -9)$,

Distance = ____

6) $(-7, -5), (5, 0)$,

Distance = ____

7) $(4, 3), (-4, -12)$,

Distance = ____

8) $(10, 1), (-5, -19)$,

Distance = ____

9) $(3, 3), (-1,5)$,

Distance = ____

10) $(2, -1), (10, 5)$,

Distance = ____

11) $(-3, 7), (-1, 4)$,

Distance = ____

12) $(5, -2), (9, -5)$,

Distance = ____

13) $(-8, 4), (4, 9)$,

Distance = ____

14) $(6, 8), (6, -6)$,

Distance = ____

15) $(9, 3), (-3, -2)$,

Distance = ____

16) $(-4, 12), (-4, 6)$,

Distance = ____

17) $(-9, -4), (-4, 8)$,

Distance = ____

18) $(11, 7), (3, 22)$,

Distance = ____

Finding Distance of Two Points - Answers

Find the distance of each pair of points.

1) $(1,9),(5,6)$,

Distance $=5$

2) $(-4,5),(8,10)$,

Distance $=13$

3) $(5,-2),(-3,4)$,

Distance $=10$

4) $(-3,0),(3,8)$,

Distance $=10$

5) $(-5,3),(4,-9)$,

Distance $=15$

6) $(-7,-5),(5,0)$,

Distance $=13$

7) $(4,3),(-4,-12)$,

Distance $=17$

8) $(10,1),(-5,-19)$,

Distance $=25$

9) $(3,3),(-1,5)$,

Distance $=\sqrt{20}=2\sqrt{5}$

10) $(2,-1),(10,5)$,

Distance $=10$

11) $(-3,7),(-1,4)$,

Distance $=\sqrt{13}$

12) $(5,-2),(9,-5)$,

Distance $=5$

13) $(-8,4),(4,9)$,

Distance $=13$

14) $(6,8),(6,-6)$,

Distance $=14$

15) $(9,3),(-3,-2)$,

Distance $=13$

16) $(-4,12),(-4,6)$,

Distance $=6$

17) $(-9,-4),(-4,8)$,

Distance $=13$

18) $(11,7),(3,22)$,

Distance $=17$

Multiplication Property of Exponents

Simplify and write the answer in exponential form.

1) $2 \times 2^2 =$

2) $5^3 \times 5 =$

3) $3^2 \times 3^2 =$

4) $4^2 \times 4^2 =$

5) $7^3 \times 7^2 \times 7 =$

6) $2 \times 2^2 \times 2^2 =$

7) $5^3 \times 5^2 \times 5 \times 5 =$

8) $2x \times x =$

9) $x^3 \times x^2 =$

10) $x^4 \times x^4 =$

11) $x^2 \times x^2 \times x^2 =$

12) $6x \times 6x =$

13) $2x^2 \times 2x^2 =$

14) $3x^2 \times x =$

15) $4x^4 \times 4x^4 \times 4x^4 =$

16) $2x^2 \times x^2 =$

17) $x^4 \times 3x =$

18) $x \times 2x^2 =$

19) $5x^4 \times 5x^4 =$

20) $2yx^2 \times 2x =$

21) $3x^4 \times y^2x^4 =$

22) $y^2x^3 \times y^5x^2 =$

23) $4yx^3 \times 2x^2y^3 =$

24) $6x^2 \times 6x^3y^4 =$

25) $3x^4y^5 \times 7x^2y^3 =$

26) $7x^2y^5 \times 9xy^3 =$

27) $7xy^4 \times 4x^3y^3 =$

28) $3x^5y^3 \times 8x^2y^3 =$

29) $3x \times y^5x^3 \times y^4 =$

30) $yx^2 \times 2y^2x^2 \times 2xy =$

31) $4yx^4 \times 5y^5x \times xy^3 =$

32) $7x^2 \times 10x^3y^3 \times 8yx^4 =$

Multiplication Property of Exponents - Answers

Simplify and write the answer in exponential form.

1) $2 \times 2^2 = 2^3$

2) $5^3 \times 5 = 5^4$

3) $3^2 \times 3^2 = 3^4$

4) $4^2 \times 4^2 = 4^4$

5) $7^3 \times 7^2 \times 7 = 7^6$

6) $2 \times 2^2 \times 2^2 = 2^5$

7) $5^3 \times 5^2 \times 5 \times 5 = 5^7$

8) $2x \times x = 2x^2$

9) $x^3 \times x^2 = x^5$

10) $x^4 \times x^4 = x^8$

11) $x^2 \times x^2 \times x^2 = x^6$

12) $6x \times 6x = 36x^2$

13) $2x^2 \times 2x^2 = 4x^4$

14) $3x^2 \times x = 3x^3$

15) $4x^4 \times 4x^4 \times 4x^4 = 64x^{12}$

16) $2x^2 \times x^2 = 2x^4$

17) $x^4 \times 3x = 3x^5$

18) $x \times 2x^2 = 2x^3$

19) $5x^4 \times 5x^4 = 25x^8$

20) $2yx^2 \times 2x = 4x^3y$

21) $3x^4 \times y^2x^4 = 3x^8y^2$

22) $y^2x^3 \times y^5x^2 = x^5y^7$

23) $4yx^3 \times 2x^2y^3 = 8x^5y^4$

24) $6x^2 \times 6x^3y^4 = 36x^5y^4$

25) $3x^4y^5 \times 7x^2y^3 = 21x^6y^8$

26) $7x^2y^5 \times 9xy^3 = 63x^3y^8$

27) $7xy^4 \times 4x^3y^3 = 28x^4y^7$

28) $3x^5y^3 \times 8x^2y^3 = 24x^7y^6$

29) $3x \times y^5x^3 \times y^4 = 3x^4y^9$

30) $yx^2 \times 2y^2x^2 \times 2xy = 4x^5y^4$

31) $4yx^4 \times 5y^5x \times xy^3 = 20x^6y^9$

32) $7x^2 \times 10x^3y^3 \times 8yx^4 = 560x^9y^4$

Division Property of Exponents

Simplify and write the answer in exponential form.

1) $\frac{2^2}{2^3} =$

2) $\frac{2^4}{2^2} =$

3) $\frac{5^5}{5} =$

4) $\frac{3}{3^5} =$

5) $\frac{x}{x^3} =$

6) $\frac{3\times3^3}{3^2\times3^4} =$

7) $\frac{5^8}{5^3} =$

8) $\frac{5\times5^6}{5^2\times5^7} =$

9) $\frac{3^4\times3^7}{3^2\times3^8} =$

10) $\frac{5x}{10x^3} =$

11) $\frac{5x^3}{2x^5} =$

12) $\frac{18x^3}{14x^6} =$

13) $\frac{12x^3}{8xy^8} =$

14) $\frac{24xy^3}{4x^4y^2} =$

15) $\frac{21x^3y^9}{7xy^5} =$

16) $\frac{36x^2y^9}{4x^3} =$

17) $\frac{12x^4y^4}{10x^6y^7} =$

18) $\frac{12y^2x^{12}}{20yx^8} =$

19) $\frac{16x^4y}{9x^8y^2} =$

20) $\frac{5x^8y^2}{20x^5y^5} =$

Division Property of Exponents - Answers

Simplify and write the answer in exponential form.

1) $\frac{2^2}{2^3} = \frac{1}{2}$

2) $\frac{2^4}{2^2} = 2^2$

3) $\frac{5^5}{5} = 5^4$

4) $\frac{3}{3^5} = \frac{1}{3^4}$

5) $\frac{x}{x^3} = \frac{1}{x^2}$

6) $\frac{3 \times 3^3}{3^2 \times 3^4} = \frac{1}{3}$

7) $\frac{5^8}{5^3} = 5^5$

8) $\frac{5 \times 5^6}{5^2 \times 5^7} = \frac{1}{5^2}$

9) $\frac{3^4 \times 3^7}{3^2 \times 3^8} = 3$

10) $\frac{5x}{10x^3} = \frac{1}{2x^2}$

11) $\frac{5x^3}{2x^5} = \frac{5}{2x^2}$

12) $\frac{18x^3}{14x^6} = \frac{9}{7x^3}$

13) $\frac{12x^3}{8xy^8} = \frac{3x^2}{2y^8}$

14) $\frac{24xy^3}{4x^4y^2} = \frac{6y}{x^3}$

15) $\frac{21x^3y^9}{7xy^5} = 3x^2y^4$

16) $\frac{36x^2y^9}{4x^3} = \frac{9y^9}{x}$

17) $\frac{12x^4y^4}{10x^6y^7} = \frac{6}{5x^2y^3}$

18) $\frac{12y^2x^{12}}{20yx^8} = \frac{3yx^4}{5}$

19) $\frac{16x^4y}{9x^8y^2} = \frac{16}{9x^4y}$

20) $\frac{5x^8y^2}{20x^5y^5} = \frac{x^3}{4y^3}$

Powers of Products and Quotients

Simplify and write the answer in exponential form.

1) $(4^2)^2 =$

2) $(6^2)^3 =$

3) $(2 \times 2^3)^4 =$

4) $(4 \times 4^4)^2 =$

5) $(3^3 \times 3^2)^3 =$

6) $(5^4 \times 5^5)^2 =$

7) $(2 \times 2^4)^2 =$

8) $(2x^6)^2 =$

9) $(11x^5)^2 =$

10) $(4x^2y^4)^4 =$

11) $(2x^4y^4)^3 =$

12) $(3x^2y^2)^2 =$

13) $(3x^4y^3)^4 =$

14) $(2x^6y^8)^2 =$

15) $(12x^3x)^3 =$

16) $(5x^9x^6)^3 =$

17) $(5x^{10}y^3)^3 =$

18) $(14x^3x^3)^2 =$

19) $(3x^3.5x)^2 =$

20) $(10x^{11}y^3)^2 =$

21) $(9x^7y^5)^2 =$

22) $(4x^4y^6)^5 =$

23) $(3x\,.4y^3)^2 =$

24) $(\frac{6x}{x^2})^2 =$

25) $\left(\frac{x^5y^5}{x^2y^2}\right)^3 =$

26) $\left(\frac{24x}{4x^6}\right)^2 =$

27) $\left(\frac{x^5}{x^7y^2}\right)^2 =$

28) $\left(\frac{xy^2}{x^2y^3}\right)^3 =$

29) $\left(\frac{4xy^4}{x^5}\right)^2 =$

30) $\left(\frac{xy^4}{5xy^2}\right)^3 =$

Powers of Products and Quotients - Answers

Simplify and write the answer in exponential form.

1) $(4^2)^2 = 4^4$

2) $(6^2)^3 = 6^6$

3) $(2 \times 2^3)^4 = 2^{16}$

4) $(4 \times 4^4)^2 = 4^{10}$

5) $(3^3 \times 3^2)^3 = 3^{15}$

6) $(5^4 \times 5^5)^2 = 5^{18}$

7) $(2 \times 2^4)^2 = 2^{10}$

8) $(2x^6)^2 = 4x^{12}$

9) $(11x^5)^2 = 121x^{10}$

10) $(4x^2y^4)^4 = 256x^8y^{16}$

11) $(2x^4y^4)^3 = 8x^{12}y^{12}$

12) $(3x^2y^2)^2 = 9x^4y^4$

13) $(3x^4y^3)^4 = 81x^{16}y^{12}$

14) $(2x^6y^8)^2 = 4x^{12}y^{16}$

15) $(12x^3x)^3 = 1{,}728x^{12}$

16) $(5x^9x^6)^3 = 125x^{45}$

17) $(5x^{10}y^3)^3 = 125x^{30}y^9$

18) $(14x^3x^3)^2 = 196x^{12}$

19) $(3x^3.5x)^2 = 225x^8$

20) $(10x^{11}y^3)^2 = 100x^{22}y^6$

21) $(9x^7y^5)^2 = 81x^{14}y^{10}$

22) $(4x^4y^6)^5 = 1{,}024x^{20}y^{30}$

23) $(3x\,.\,4y^3)^2 = 144x^2y^6$

24) $(\frac{6x}{x^2})^2 = \frac{36}{x^2}$

25) $\left(\frac{x^5y^5}{x^2y^2}\right)^3 = x^9y^9$

26) $\left(\frac{24x}{4x^6}\right)^2 = \frac{36}{x^{10}}$

27) $\left(\frac{x^5}{x^7y^2}\right)^2 = \frac{1}{x^4y^4}$

28) $\left(\frac{xy^2}{x^2y^3}\right)^3 = \frac{1}{x^3y^3}$

29) $\left(\frac{4xy^4}{x^5}\right)^2 = \frac{16y^8}{x^8}$

30) $\left(\frac{xy^4}{5xy^2}\right)^3 = \frac{y^6}{125}$

Zero and Negative Exponents

Evaluate the following expressions.

1) $1^{-1} =$

2) $2^{-2} =$

3) $0^{15} =$

4) $1^{-10} =$

5) $8^{-1} =$

6) $8^{-2} =$

7) $2^{-4} =$

8) $10^{-2} =$

9) $9^{-2} =$

10) $3^{-3} =$

11) $7^{-3} =$

12) $3^{-4} =$

13) $6^{-3} =$

14) $5^{-3} =$

15) $22^{-1} =$

16) $4^{-4} =$

17) $5^{-4} =$

18) $15^{-2} =$

19) $4^{-5} =$

20) $9^{-3} =$

21) $3^{-5} =$

22) $5^{-4} =$

23) $12^{-3} =$

24) $15^{-3} =$

25) $20^{-3} =$

26) $50^{-2} =$

27) $18^{-3} =$

28) $24^{-2} =$

29) $30^{-3} =$

30) $10^{-5} =$

31) $(\frac{1}{8})^{-1}$

32) $(\frac{1}{5})^{-2} =$

33) $(\frac{1}{7})^{-2} =$

34) $(\frac{2}{3})^{-2} =$

35) $(\frac{1}{5})^{-3} =$

36) $(\frac{3}{4})^{-2} =$

37) $(\frac{2}{5})^{-2} =$

38) $(\frac{1}{2})^{-8} =$

39) $(\frac{2}{5})^{-3} =$

40) $(\frac{3}{7})^{-2} =$

41) $(\frac{5}{6})^{-3} =$

42) $(\frac{4}{9})^{-2} =$

bit.ly/3mkh4v

Find more at

Zero and Negative Exponents - Answers

Evaluate the following expressions.

1) $1^{-1} = 1$

2) $2^{-2} = \frac{1}{4}$

3) $0^{15} = 0$

4) $1^{-10} = 1$

5) $8^{-1} = \frac{1}{8}$

6) $8^{-2} = \frac{1}{64}$

7) $2^{-4} = \frac{1}{16}$

8) $10^{-2} = \frac{1}{100}$

9) $9^{-2} = \frac{1}{81}$

10) $3^{-3} = \frac{1}{27}$

11) $7^{-3} = \frac{1}{343}$

12) $3^{-4} = \frac{1}{81}$

13) $6^{-3} = \frac{1}{216}$

14) $5^{-3} = \frac{1}{125}$

15) $22^{-1} = \frac{1}{22}$

16) $4^{-4} = \frac{1}{256}$

17) $5^{-4} = \frac{1}{625}$

18) $15^{-2} = \frac{1}{225}$

19) $4^{-5} = \frac{1}{1,024}$

20) $9^{-3} = \frac{1}{729}$

21) $3^{-5} = \frac{1}{243}$

22) $5^{-4} = \frac{1}{625}$

23) $12^{-2} = \frac{1}{144}$

24) $15^{-3} = \frac{1}{3,375}$

25) $20^{-3} = \frac{1}{8,000}$

26) $50^{-2} = \frac{1}{2,500}$

27) $18^{-3} = \frac{1}{5,832}$

28) $24^{-2} = \frac{1}{576}$

29) $30^{-3} = \frac{1}{27,000}$

30) $10^{-5} = \frac{1}{100,000}$

31) $(\frac{1}{8})^{-1} = 8$

32) $(\frac{1}{5})^{-2} = 25$

33) $(\frac{1}{7})^{-2} = 49$

34) $(\frac{2}{3})^{-2} = \frac{9}{4}$

35) $(\frac{1}{5})^{-3} = 125$

36) $(\frac{3}{4})^{-2} = \frac{64}{27}$

37) $(\frac{2}{5})^{-2} = \frac{25}{4}$

38) $(\frac{1}{2})^{-8} = 256$

39) $(\frac{2}{5})^{-3} = \frac{125}{8}$

40) $(\frac{3}{7})^{-2} = \frac{49}{9}$

41) $(\frac{5}{6})^{-3} = \frac{216}{125}$

42) $(\frac{4}{9})^{-2} = \frac{81}{16}$

Negative Exponents and Negative Bases

Simplify and write the answer.

1) $-3^{-1} =$

2) $-5^{-2} =$

3) $-2^{-4} =$

4) $-x^{-3} =$

5) $2x^{-1} =$

6) $-4x^{-3} =$

7) $-12x^{-5} =$

8) $-5x^{-2}y^{-3} =$

9) $20x^{-4}y^{-1} =$

10) $14a^{-6}b^{-7} =$

11) $-12x^{2}y^{-3} =$

12) $-\frac{25}{x^{-6}} =$

13) $-\frac{2x}{a^{-4}} =$

14) $(-\frac{1}{3x})^{-2} =$

15) $(-\frac{3}{4x})^{-2} =$

16) $-\frac{9}{a^{-7}b^{-2}} =$

17) $-\frac{5x}{x^{-3}} =$

18) $-\frac{a^{-3}}{b^{-2}} =$

19) $-\frac{8}{x^{-3}} =$

20) $\frac{5b}{-9c^{-4}} =$

21) $\frac{9ab}{a^{-3}b^{-1}} =$

22) $-\frac{15a^{-2}}{30b^{-3}} =$

23) $\frac{4ab^{-2}}{-3c^{-2}} =$

24) $(\frac{3a}{2c})^{-2} =$

25) $(-\frac{5x}{3yz})^{-3} =$

26) $\frac{11ab^{-2}}{-3c^{-2}} =$

27) $(-\frac{x^{3}}{x^{4}})^{-2} =$

28) $(-\frac{x^{-2}}{3x^{2}})^{-3} =$

Negative Exponents and Negative Bases - Answers

Simplify and write the answer.

1) $-3^{-1} = -\frac{1}{3}$
2) $-5^{-2} = -\frac{1}{25}$
3) $-2^{-4} = -\frac{1}{16}$
4) $-x^{-3} = -\frac{1}{x^3}$
5) $2x^{-1} = \frac{2}{x}$
6) $-4x^{-3} = -\frac{4}{x^3}$
7) $-12x^{-5} = -\frac{12}{x^5}$
8) $-5x^{-2}y^{-3} = -\frac{5}{x^2y^3}$
9) $20x^{-4}y^{-1} = \frac{20}{x^4y}$
10) $14a^{-6}b^{-7} = \frac{14}{a^6b^7}$
11) $-12x^2y^{-3} = -\frac{12x^2}{y^3}$
12) $-\frac{25}{x^{-6}} = -25x^6$
13) $-\frac{2x}{a^{-4}} = -2xa^4$
14) $(-\frac{1}{3x})^{-2} = 9x^2$
15) $(-\frac{3}{4x})^{-2} = \frac{16x^2}{9}$
16) $-\frac{9}{a^{-7}b^{-2}} = -9a^7b^2$
17) $-\frac{5x}{x^{-3}} = -5x^4$
18) $-\frac{a^{-3}}{b^{-2}} = -\frac{b^2}{a^3}$
19) $-\frac{8}{x^{-3}} = -8x^3$
20) $\frac{5b}{-9c^{-4}} = -\frac{5bc^4}{9}$
21) $\frac{9ab}{a^{-3}b^{-1}} = 9a^4b^2$
22) $-\frac{15a^{-2}}{30b^{-3}} = -\frac{b^3}{2a^2}$
23) $\frac{4ab^{-2}}{-3c^{-2}} = -\frac{4ac^2}{3b^2}$
24) $(\frac{3a}{2c})^{-2} = \frac{4c^2}{9a^2}$
25) $(-\frac{5x}{3yz})^{-3} = -\frac{27y^3z^3}{125x^3}$
26) $\frac{11ab^{-2}}{-3c^{-2}} = -\frac{11ac^2}{3b^2}$
27) $\left(-\frac{x^3}{x^4}\right)^{-2} = x^2$
28) $(-\frac{x^{-2}}{3x^2})^{-3} = -27x^{12}$

Scientific Notation

Write each number in scientific notation.

1) 0.113 =
2) 0.02 =
3) 7.5 =
4) 20 =
5) 60 =
6) 0.004 =
7) 78 =
8) 1,600 =
9) 1,450 =
10) 31,000 =
11) 2,000,000 =
12) 0.0000003 =
13) 554,000 =
14) 0.000725 =
15) 0.00034 =
16) 86,000,000 =
17) 62,000 =
18) 97,000,000 =
19) 0.0000045 =
20) 0.0019 =

Write each number in standard notation.

1) $2 \times 10^{-1} =$
2) $8 \times 10^{-2} =$
3) $1.8 \times 10^{3} =$
4) $9 \times 10^{-4} =$
5) $1.7 \times 10^{-2} =$
6) $9 \times 10^{3} =$
7) $7 \times 10^{5} =$
8) $1.15 \times 10^{4} =$
9) $7 \times 10^{-5} =$
10) $8.3 \times 10^{-5} =$

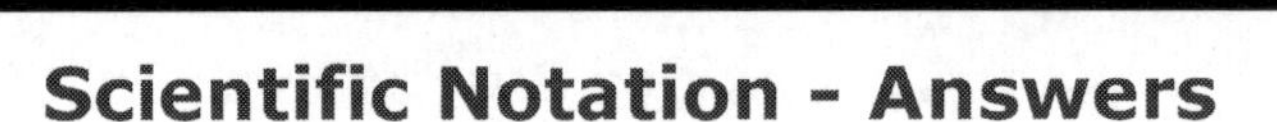

Scientific Notation - Answers

Write each number in scientific notation.

1) $0.113 = 1.13 \times 10^{-1}$
2) $0.02 = 2 \times 10^{-2}$
3) $7.5 = 2.5 \times 10^{0}$
4) $20 = 2 \times 10^{1}$
5) $60 = 6 \times 10^{1}$
6) $0.004 = 4 \times 10^{-3}$
7) $78 = 7.8 \times 10^{1}$
8) $1{,}600 = 1.6 \times 10^{3}$
9) $1{,}450 = 1.45 \times 10^{3}$
10) $31{,}000 = 3.1 \times 10^{4}$
11) $2{,}000{,}000 = 2 \times 10^{6}$
12) $0.0000003 = 3 \times 10^{-7}$
13) $554{,}000 = 5.54 \times 10^{5}$
14) $0.000725 = 7.25 \times 10^{-4}$
15) $0.00034 = 3.4 \times 10^{-4}$
16) $86{,}000{,}000 = 8.6 \times 10^{7}$
17) $62{,}000 = 6.2 \times 10^{4}$
18) $97{,}000{,}000 = 9.7 \times 10^{7}$
19) $0.0000045 = 4.5 \times 10^{-6}$
20) $0.0019 = 1.9 \times 10^{-3}$

Write each number in standard notation.

1) $2 \times 10^{-1} = 0.2$
2) $8 \times 10^{-2} = 0.08$
3) $1.8 \times 10^{3} = 1{,}800$
4) $9 \times 10^{-4} = 0.0009$
5) $1.7 \times 10^{-2} = 0.017$
6) $9 \times 10^{3} = 9{,}000$
7) $7 \times 10^{5} = 700{,}000$
8) $1.15 \times 10^{4} = 11{,}500$
9) $7 \times 10^{-5} = 0.00007$
10) $8.3 \times 10^{-5} = 0.000083$

Radicals

Simplify and write the answer.

1) $\sqrt{0} =$ ____
2) $\sqrt{1} =$ ____
3) $\sqrt{4} =$ ____
4) $\sqrt{16} =$ ____
5) $\sqrt{9} =$ ____
6) $\sqrt{25} =$ ____
7) $\sqrt{49} =$ ____
8) $\sqrt{36} =$ ____
9) $\sqrt{64} =$ ____
10) $\sqrt{81} =$ ____
11) $\sqrt{121} =$ ____
12) $\sqrt{225} =$ ____
13) $\sqrt{144} =$ ____
14) $\sqrt{100} =$ ____
15) $\sqrt{256} =$ ____
16) $\sqrt{289} =$ ____
17) $\sqrt{324} =$ ____
18) $\sqrt{400} =$ ____
19) $\sqrt{900} =$ ____
20) $\sqrt{529} =$ ____
21) $\sqrt{361} =$ ____
22) $\sqrt{169} =$ ____
23) $\sqrt{196} =$ ____
24) $\sqrt{90} =$ ____

Evaluate.

1) $\sqrt{6} \times \sqrt{6} =$
2) $\sqrt{5} \times \sqrt{5} =$
3) $\sqrt{8} \times \sqrt{8} =$
4) $\sqrt{2} + \sqrt{2} =$
5) $\sqrt{8} + \sqrt{8} =$
6) $6\sqrt{5} - 2\sqrt{5} =$
7) $\sqrt{25} \times \sqrt{16} =$
8) $\sqrt{25} \times \sqrt{64} =$
9) $\sqrt{81} \times \sqrt{25} =$
10) $5\sqrt{3} \times 2\sqrt{3} =$
11) $8\sqrt{2} \times 2\sqrt{2} =$
12) $6\sqrt{3} - \sqrt{12} =$

Radicals - Answers

Simplify and write the answer.

1) $\sqrt{0} = 0$
2) $\sqrt{1} = 1$
3) $\sqrt{4} = 2$
4) $\sqrt{16} = 4$
5) $\sqrt{9} = 3$
6) $\sqrt{25} = 5$
7) $\sqrt{49} = 7$
8) $\sqrt{36} = 6$
9) $\sqrt{64} = 8$
10) $\sqrt{81} = 9$
11) $\sqrt{121} = 11$
12) $\sqrt{225} = 15$
13) $\sqrt{144} = 12$
14) $\sqrt{100} = 10$
15) $\sqrt{256} = 16$
16) $\sqrt{289} = 17$
17) $\sqrt{324} = 18$
18) $\sqrt{400} = 20$
19) $\sqrt{900} = 30$
20) $\sqrt{529} = 23$
21) $\sqrt{361} = 19$
22) $\sqrt{169} = 13$
23) $\sqrt{196} = 14$
24) $\sqrt{90} = 3\sqrt{10}$

Evaluate.

1) $\sqrt{6} \times \sqrt{6} = 6$
2) $\sqrt{5} \times \sqrt{5} = 5$
3) $\sqrt{8} \times \sqrt{8} = 8$
4) $\sqrt{2} + \sqrt{2} = 2\sqrt{2}$
5) $\sqrt{8} + \sqrt{8} = 2\sqrt{8} = 4\sqrt{2}$
6) $6\sqrt{5} - 2\sqrt{5} = 4\sqrt{5}$
7) $\sqrt{25} \times \sqrt{16} = 20$
8) $\sqrt{25} \times \sqrt{64} = 40$
9) $\sqrt{81} \times \sqrt{25} = 45$
10) $5\sqrt{3} \times 2\sqrt{3} = 30$
11) $8\sqrt{2} \times 2\sqrt{2} = 32$
12) $6\sqrt{3} - \sqrt{12} = 4\sqrt{3}$

Simplifying Polynomials

Simplify each expression.

1) $2(2x + 2) =$ ____________

2) $4(4x - 2) =$ ____________

3) $3(5x + 3) =$ ____________

4) $6(7x + 5) =$ ____________

5) $-3(8x - 7) =$ ____________

6) $2x(3x + 4) =$ ____________

7) $3x^2 + 3x^2 - 2x^3 =$ ____________

8) $2x - x^2 + 6x^3 + 4 =$ ____________

9) $5x + 2x^2 - 9x^3 =$ ____________

10) $7x^2 + 5x^4 - 2x^3 =$ ____________

11) $-3x^2 + 5x^3 + 6x^4 =$ ____________

12) $(x - 3)(x - 4) =$ ____________

13) $(x - 5)(x + 4) =$ ____________

14) $(x - 6)(x - 3) =$ ____________

15) $(2x + 5)(x + 8) =$ ____________

16) $(3x - 8)(x + 4) =$ ____________

17) $-8x^2 + 2x^3 - 10x^4 + 5x =$ ____________

18) $11 - 6x^2 + 5x^2 - 12x^3 + 22 =$ ____________

19) $2x^2 - 2x + 3x^3 + 12x - 22x =$ ____________

20) $11 - 4x^2 + 3x^2 - 7x^3 + 3 =$ ____________

21) $2x^5 - x^3 + 8x^2 - 2x^5 =$ ____________

22) $(2x^3 - 1) + (3x^3 - 2x^3) =$ ____________

Simplifying Polynomials - Answers

Simplify each expression.

1) $2(2x + 2) =$

$4x + 4$

2) $4(4x - 2) =$

$16x - 8$

3) $3(5x + 3) =$

$15x + 9$

4) $6(7x + 5) =$

$42x + 30$

5) $-3(8x - 7) =$

$-24x + 21$

6) $2x(3x + 4) =$

$6x^2 + 8x$

7) $3x^2 + 3x^2 - 2x^3 =$

$-2x^3 + 6x^2$

8) $2x - x^2 + 6x^3 + 4 =$

$6x^3 - x^2 + 2x + 4$

9) $5x + 2x^2 - 9x^3 =$

$-9x^3 + 2x^2 + 5x$

10) $7x^2 + 5x^4 - 2x^3 =$

$5x^4 - 2x^3 + 7x^2$

11) $-3x^2 + 5x^3 + 6x^4 =$

$6x^4 + 5x^3 - 3x^2$

12) $(x - 3)(x - 4) =$

$x^2 - 7x + 12$

13) $(x - 5)(x + 4) =$

$x^2 - x - 20$

14) $(x - 6)(x - 3) =$

$x^2 - 9x + 18$

15) $(2x + 5)(x + 8) =$

$2x^2 + 21x + 40$

16) $(3x - 8)(x + 4) =$

$3x^2 + 4x - 32$

17) $-8x^2 + 2x^3 - 10x^4 + 5x =$

$-10x^4 + 2x^3 - 8x^2 + 5x$

18) $11 - 6x^2 + 5x^2 - 12x^3 + 22 =$

$-12x^3 - x^2 + 33$

19) $2x^2 - 2x + 3x^3 + 12x - 22x =$

$3x^3 + 2x^2 - 12x$

20) $11 - 4x^2 + 3x^2 - 7x^3 + 3 =$

$-7x^3 - x^2 + 14$

21) $2x^5 - x^3 + 8x^2 - 2x^5 =$

$-x^3 + 8x^2$

22) $(2x^3 - 1) + (3x^3 - 2x^3) =$

$3x^3 - 1$

Adding and Subtracting Polynomials

Add or subtract expressions.

1) $(x^2 - 3) + (x^2 + 1) =$ ________

2) $(2x^2 - 4) - (2 - 4x^2) =$ ________

3) $(x^3 + 2x^2) - (x^3 + 5) =$ ________

4) $(3x^3 - x^2) + (4x^2 - 7x) =$ ________

5) $(2x^3 + 3x) - (5x^3 + 2) =$ ________

6) $(5x^3 - 2) + (2x^3 + 10) =$ ________

7) $(7x^3 + 5) - (9 - 4x^3) =$ ________

8) $(5x^2 + 3x^3) - (2x^3 + 6) =$ ________

9) $(8x^2 - x) + (4x - 8x^2) =$ ________

10) $(6x + 9x^2) - (5x + 2) =$ ________

11) $(7x^4 - 2x) - (6x - 2x^4) =$ ________

12) $(2x - 4x^3) - (9x^3 + 6x) =$ ________

13) $(8x^3 - 8x^2) - (6x^2 - 3x) =$ ________

14) $(9x^2 - 6) + (5x^2 - 4x^3) =$ ________

15) $(8x^3 + 3x^4) - (x^4 - 3x^3) =$ ________

16) $(-4x^3 - 2x) + (5x - 2x^3) =$ ________

17) $(6x - 4x^4) - (8x^4 + 3x) =$ ________

18) $(7x - 8x^2) - (9x^4 - 3x^2) =$ ________

19) $(9x^3 - 6) + (9x^3 - 5x^2) =$ ________

20) $(5x^3 + x^4) - (8x^4 - 7x^3) =$ ________

Adding and Subtracting Polynomials - Answers

Add or subtract expressions.

1) $(x^2 - 3) + (x^2 + 1) =$

$2x^2 - 2$

2) $(2x^2 - 4) - (2 - 4x^2) =$

$6x^2 - 6$

3) $(x^3 + 2x^2) - (x^3 + 5) =$

$2x^2 - 5$

4) $(3x^3 - x^2) + (4x^2 - 7x) =$

$3x^3 + 3x^2 - 7x$

5) $(2x^3 + 3x) - (5x^3 + 2) =$

$-3x^3 + 3x^2 - 2$

6) $(5x^3 - 2) + (2x^3 + 10) =$

$7x^3 + 8$

7) $(7x^3 + 5) - (9 - 4x^3) =$

$11x^3 - 4$

8) $(5x^2 + 3x^3) - (2x^3 + 6) =$

$x^3 + 5x^2 - 6$

9) $(8x^2 - x) + (4x - 8x^2) =$

$3x$

10) $(6x + 9x^2) - (5x + 2) =$

$9x^2 + x - 2$

11) $(7x^4 - 2x) - (6x - 2x^4) =$

$9x^4 - 8x$

12) $(2x - 4x^3) - (9x^3 + 6x) =$

$-13x^3 - 4x$

13) $(8x^3 - 8x^2) - (6x^2 - 3x) =$

$8x^3 - 14x^2 + 3x$

14) $(9x^2 - 6) + (5x^2 - 4x^3) =$

$-4x^3 + 14x^2 - 6$

15) $(8x^3 + 3x^4) - (x^4 - 3x^3) =$

$2x^4 + 11x^3$

16) $(-4x^3 - 2x) + (5x - 2x^3) =$

$-6x^3 + 3x$

17) $(6x - 4x^4) - (8x^4 + 3x) =$

$-12x^4 + 3x$

18) $(7x - 8x^2) - (9x^4 - 3x^2) =$

$-9x^4 - 5x^2 + 7x$

19) $(9x^3 - 6) + (9x^3 - 5x^2) =$

$18x^3 - 5x^2 - 6$

20) $(5x^3 + x^4) - (8x^4 - 7x^3) =$

$-7x^4 + 12x^3$

Multiplying Monomials

Simplify each expression.

1) $5x^8 \times x^3 =$ ____________

2) $-4z^7 \times 5z^5 =$ ____________

3) $-6xy^8 \times 3x^5y^3 =$ ____________

4) $5xy^5 \times 3x^3y^4 =$ ____________

5) $8s^6t^2 \times 6s^3t^7 =$ ____________

6) $9xy^6z \times 3y^4z^2 =$ ____________

7) $4pq^5 \times (-7p^4q^8) =$ ____________

8) $10p^3q^5 \times (-4p^4q^6) =$ ____________

9) $(-9a^4b^7c^4) \times (-4a^7b) =$ ____________

10) $5u^3v^9z^2 \times (-4uv^9z) =$ ____________

11) $8x^3y^2z^5 \times (-9x^4y^2z) =$ ____________

12) $5y^5 \times 6y^3 =$ ____________

13) $7x^5y \times 3xy^2 =$ ____________

14) $7a^4b^2 \times 3a^8b =$ ____________

15) $5p^5q^4 \times (-6pq^4) =$ ____________

16) $(-8x^5y^2) \times 4x^6y^3 =$ ____________

17) $12x^5y^4 \times 2x^8y =$ ____________

18) $9s^4t^2 \times (-5st^5) =$ ____________

19) $(-5p^2q^4r) \times 7pq^5r^3 =$ ____________

20) $7u^5v^9 \times (-5u^{12}v^7) =$ ____________

21) $(-9xy^2z^4) \times 2x^2yz^5 =$ ____________

22) $6a^8b^8c^{12} \times 9a^7b^5c^8 =$ ____________

Multiplying Monomials - Answers

Simplify each expression.

1) $5x^8 \times x^3 =$
$5x^{11}$

2) $-4z^7 \times 5z^5 =$
$-20z^{12}$

3) $-6xy^8 \times 3x^5y^3 =$
$-18x^6y^{11}$

4) $5xy^5 \times 3x^3y^4 =$
$15x^4y^9$

5) $8s^6t^2 \times 6s^3t^7 =$
$48s^9t^9$

6) $9xy^6z \times 3y^4z^2 =$
$27xy^{10}z^3$

7) $4pq^5 \times (-7p^4q^8) =$
$-28p^5q^{13}$

8) $10p^3q^5 \times (-4p^4q^6) =$
$-40p^7q^{11}$

9) $(-9a^4b^7c^4) \times (-\ 4a^7b) =$
$36a^{11}b^8c^4$

10) $5u^3v^9z^2 \times (-4uv^9z) =$
$-20u^4v^{18}z^3$

11) $8x^3y^2z^5 \times (-\ 9x^4y^2z) =$
$-\ 72x^7y^4z^6$

12) $5y^5 \times 6y^3 =$
$30y^8$

13) $7x^5y \times 3xy^2 =$
$21x^6y^3$

14) $7a^4b^2 \times 3a^8b =$
$21a^{12}b^3$

15) $5p^5q^4 \times (-6pq^4) =$
$-30p^6q^8$

16) $(-8x^5y^2) \times 4x^6y^3 =$
$-32x^{11}y^5$

17) $12x^5y^4 \times 2x^8y =$
$24x^{13}y^5$

18) $9s^4t^2 \times (-5st^5) =$
$-45s^5t^7$

19) $(-5p^2q^4r) \times 7pq^5r^3 =$
$-35p^3q^9r^4$

20) $7u^5v^9 \times (-5u^{12}v^7) =$
$-35u^{17}v^{16}$

21) $(-9xy^2z^4) \times 2x^2yz^5 =$
$-18x^3y^3z^9$

22) $6a^8b^8c^{12} \times 9a^7b^5c^8 =$
$54a^{15}b^{13}c^{20}$

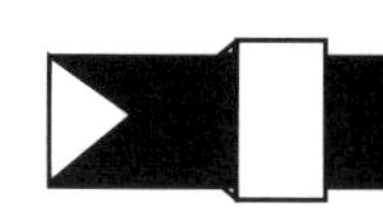

Multiplying and Dividing Monomials

Simplify each expression.

1) $(8x^3)(2x^2) =$ ____

2) $(4x^6)(5x^4) =$ ____

3) $(-6x^8)(3x^3) =$ ____

4) $(5x^8y^9)(-6x^6y^9) =$ ____

5) $(8x^5y^6)(3x^2y^5) =$ ____

6) $(8yx^2)(7y^5x^3) =$ ____

7) $(4x^2y)(2x^2y^3) =$ ____

8) $(-2x^9y^4)(-9x^6y^8) =$ ____

9) $(-5x^8y^2)(-6x^4y^5) =$ ____

10) $(8x^8y)(-7x^4y^3) =$ ____

11) $(9x^6y^2)(6x^7y^4) =$ ____

12) $(8x^9y^5)(6x^5y^4) =$ ____

13) $(-5x^8y^9)(7x^7y^8) =$ ____

14) $(6x^2y^5)(5x^3y^2) =$ ____

15) $(9x^5y^{12})(4x^7y^9) =$ ____

16) $(-10x^{14}y^8)(2x^7y^5) =$ ____

17) $\frac{8x^4y^3}{xy^2} =$ ____

18) $\frac{6x^5y^6}{2x^3y} =$ ____

19) $\frac{12x^3y^7}{4xy} =$ ____

20) $\frac{-20x^8y^9}{5x^5y^4} =$ ____

Multiplying and Dividing Monomials - Answers

Simplify each expression.

1) $(8x^3)(2x^2) =$

$16x^5$

2) $(4x^6)(5x^4) =$

$20x^{10}$

3) $(-6x^8)(3x^3) =$

$-18x^{11}$

4) $(5x^8y^9)(-6x^6y^9) =$

$-30x^{14}y^{18}$

5) $(8x^5y^6)(3x^2y^5) =$

$24x^7y^{11}$

6) $(8yx^2)(7y^5x^3) =$

$56y^6x^5$

7) $(4x^2y)(2x^2y^3) =$

$8x^4y^4$

8) $(-2x^9y^4)(-9x^6y^8) =$

$18x^{15}y^{12}$

9) $(-5x^8y^2)(-6x^4y^5) =$

$30x^{12}y^7$

10) $(8x^8y)(-7x^4y^3) =$

$-56x^{12}y^4$

11) $(9x^6y^2)(6x^7y^4) =$

$54x^{13}y^6$

12) $(8x^9y^5)(6x^5y^4) =$

$48x^{14}y^9$

13) $(-5x^8y^9)(7x^7y^8) =$

$-35x^{15}y^{17}$

14) $(6x^2y^5)(5x^3y^2) =$

$30x^5y^7$

15) $(9x^5y^{12})(4x^7y^9) =$

$36x^{12}y^{21}$

16) $(-10x^{14}y^8)(2x^7y^5) =$

$-20x^{21}y^{13}$

17) $\frac{8x^4y^3}{xy^2} =$

$8x^3y$

18) $\frac{6x^5y^6}{2x^3y} =$

$3x^2y^5$

19) $\frac{12x^3y^7}{4xy} =$

$3x^2y^6$

20) $\frac{-20x^8y^9}{5x^5y^4} =$

$-4x^3y^5$

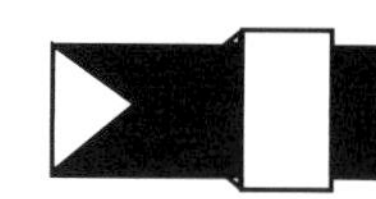

Multiplying a Polynomial and a Monomial

Find each product.

1) $x(x-2) =$ ________

2) $2(2+x) =$ ________

3) $x(x-1) =$ ________

4) $x(x+3) =$ ________

5) $2x(x-2) =$ ________

6) $5(4x+3) =$ ________

7) $4x(3x-4) =$ ________

8) $x(5x+2y) =$ ________

9) $3x(x-2y) =$ ________

10) $6x(3x-4y) =$ ________

11) $2x(3x-8) =$ ________

12) $6x(4x-6y) =$ ________

13) $3x(4x-2y) =$ ________

14) $2x(2x-6y) =$ ________

15) $5x(x^2+y^2) =$ ________

16) $3x(2x^2-y^2) =$ ________

17) $7(2x^2+9y^2) =$ ________

18) $2x(-2x^2y+3y) =$ ________

19) $-2(2x^2-4xy+2) =$ ________

20) $5(x^2-6xy-8) =$ ________

Multiplying a Polynomial and a Monomial - Answers

Find each product.

1) $x(x-2) =$

$x^2 - 2x$

2) $2(2+x) =$

$2x + 4$

3) $x(x-1) =$

$x^2 - x$

4) $x(x+3) =$

$x^2 + 3x$

5) $2x(x-2) =$

$2x^2 - 4x$

6) $5(4x+3) =$

$20x + 15$

7) $4x(3x-4) =$

$12x^2 - 16x$

8) $x(5x+2y) =$

$5x^2 + 2xy$

9) $3x(x-2y) =$

$3x^2 - 6xy$

10) $6x(3x-4y) =$

$18x^2 - 24xy$

11) $2x(3x-8) =$

$6x^2 - 16x$

12) $6x(4x-6y) =$

$24x^2 - 36xy$

13) $3x(4x-2y) =$

$12x^2 - 6xy$

14) $2x(2x-6y) =$

$4x^2 - 12xy$

15) $5x(x^2+y^2) =$

$5x^3 + 5xy^2$

16) $3x(2x^2-y^2) =$

$6x^3 - 3xy^2$

17) $7(2x^2+9y^2) =$

$14x^2 + 63y^2$

18) $2x(-2x^2y+3y) =$

$-4x^3y + 6xy$

19) $-2(2x^2-4xy+2) =$

$-4x^2 + 8xy - 4$

20) $5(x^2-6xy-8) =$

$5x^2 - 30xy - 40$

Multiplying Binomials

Find each product.

1) $(x-2)(x+5) =$ ________

2) $(x+4)(x+2) =$ ________

3) $(x-2)(x-4) =$ ________

4) $(x-8)(x-2) =$ ________

5) $(x-7)(x-5) =$ ________

6) $(x+6)(x+2) =$ ________

7) $(x-9)(x+3) =$ ________

8) $(x-8)(x-5) =$ ________

9) $(x+3)(x+7) =$ ________

10) $(x-9)(x+4) =$ ________

11) $(x+6)(x+6) =$ ________

12) $(x+7)(x+7) =$ ________

13) $(x-8)(x+7) =$ ________

14) $(x+9)(x+9) =$ ________

15) $(x-8)(x-8) =$ ________

16) $(2x-9)(x+5) =$ ________

17) $(2x-3)(x+4) =$ ________

18) $(2x+4)(x+2) =$ ________

19) $(2x+2)(x+3) =$ ________

20) $(2x-4)(2x+2) =$ ________

Multiplying Binomials - Answers

Find each product.

1) $(x-2)(x+5) =$

$x^2+3x-10$

2) $(x+4)(x+2) =$

x^2+6x+8

3) $(x-2)(x-4) =$

x^2-6x+8

4) $(x-8)(x-2) =$

$x^2-10x+16$

5) $(x-7)(x-5) =$

$x^2-12x+35$

6) $(x+6)(x+2) =$

$x^2+8x+12$

7) $(x-9)(x+3) =$

$x^2-6x-27$

8) $(x-8)(x-5) =$

$x^2-13x+40$

9) $(x+3)(x+7) =$

$x^2+10x+21$

10) $(x-9)(x+4) =$

$x^2-5x-36$

11) $(x+6)(x+6) =$

$x^2+12x+36$

12) $(x+7)(x+7) =$

$x^2+14x+49$

13) $(x-8)(x+7) =$

x^2-x-56

14) $(x+9)(x+9) =$

$x^2+18x+81$

15) $(x-8)(x-8) =$

$x^2-16x+64$

16) $(2x-9)(x+5) =$

$2x^2+x-45$

17) $(2x-3)(x+4) =$

$2x^2+5x-12$

18) $(2x+4)(x+2) =$

$2x^2+8x+8$

19) $(2x+2)(x+3) =$

$2x^2+8x+6$

20) $(2x-4)(2x+2) =$

$4x^2-4x-8$

Factoring Trinomials

Factor each trinomial.

1) $x^2 + 3x - 10 =$

2) $x^2 + 6x + 8 =$

3) $x^2 - 6x + 8 =$

4) $x^2 - 10x + 16 =$

5) $x^2 - 13x + 40 =$

6) $x^2 + 8x + 12 =$

7) $x^2 - 6x - 27 =$

8) $x^2 - 14x + 48 =$

9) $x^2 + 15x + 56 =$

10) $x^2 - 5x - 36 =$

11) $x^2 + 12x + 36 =$

12) $x^2 + 16x + 63 =$

13) $x^2 + x - 72 =$

14) $x^2 + 18x + 81 =$

15) $x^2 - 16x + 64 =$

16) $x^2 - 18x + 81 =$

17) $2x^2 + 8x + 6 =$

18) $2x^2 + 6x - 8 =$

19) $2x^2 + 12x + 10 =$

20) $4x^2 + 6x - 28 =$

Factoring Trinomials – Answer

Factor each trinomial.

1) $x^2 + 3x - 10 =$

$(x - 2)(x + 5)$

2) $x^2 + 6x + 8 =$

$(x + 4)(x + 2)$

3) $x^2 - 6x + 8 =$

$(x - 2)(x - 4)$

4) $x^2 - 10x + 16 =$

$(x - 8)(x - 2)$

5) $x^2 - 13x + 40 =$

$(x - 8)(x - 5)$

6) $x^2 + 8x + 12 =$

$(x + 6)(x + 2)$

7) $x^2 - 6x - 27 =$

$(x - 9)(x + 3)$

8) $x^2 - 14x + 48 =$

$(x - 8)(x - 6)$

9) $x^2 + 15x + 56 =$

$(x + 8)(x + 7)$

10) $x^2 - 5x - 36 =$

$(x - 9)(x + 4)$

11) $x^2 + 12x + 36 =$

$(x + 6)(x + 6)$

12) $x^2 + 16x + 63 =$

$(x + 7)(x + 9)$

13) $x^2 + x - 72 =$

$(x - 8)(x + 9)$

14) $x^2 + 18x + 81 =$

$(x + 9)(x + 9)$

15) $x^2 - 16x + 64 =$

$(x - 8)(x - 8)$

16) $x^2 - 18x + 81 =$

$(x - 9)(x - 9)$

17) $2x^2 + 8x + 6 =$

$(2x + 2)(x + 3)$

18) $2x^2 + 6x - 8 =$

$(2x - 2)(x + 4)$

19) $2x^2 + 12x + 10 =$

$(2x + 2)(x + 5)$

20) $4x^2 + 6x - 28 =$

$(2x - 4)(2x + 7)$

The Pythagorean Theorem

Do the following lengths form a right triangle?

1) _____

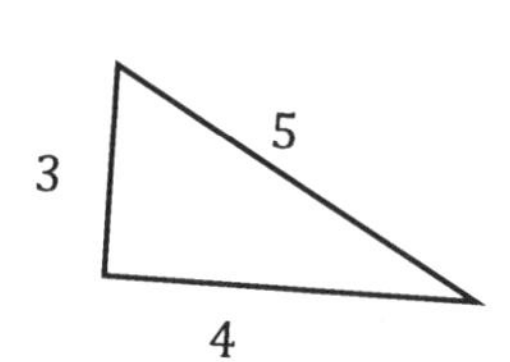

2) _____

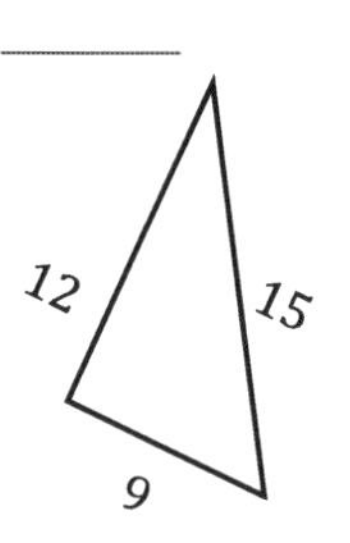

3) _____

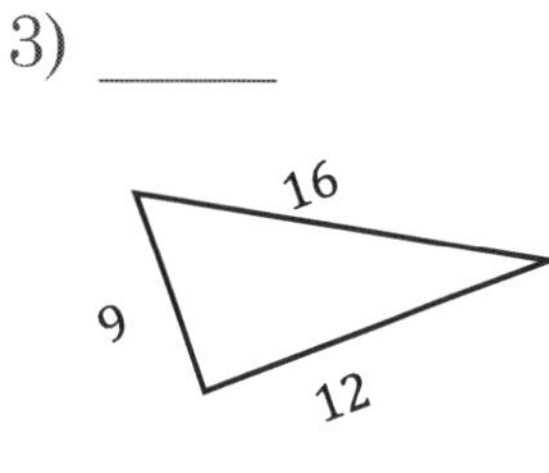

4) _____

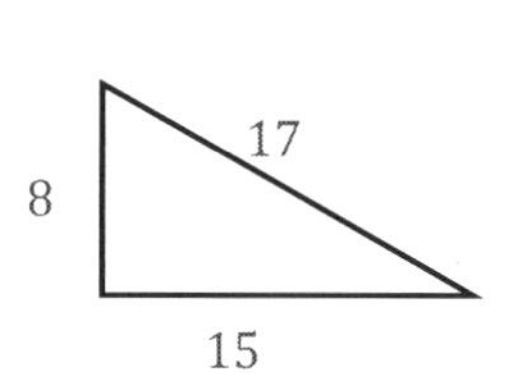

5) _____

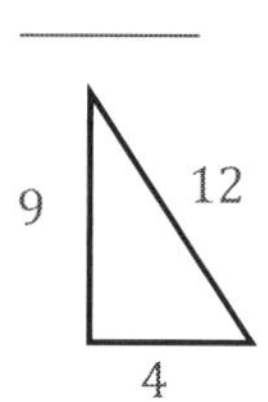

6) _____

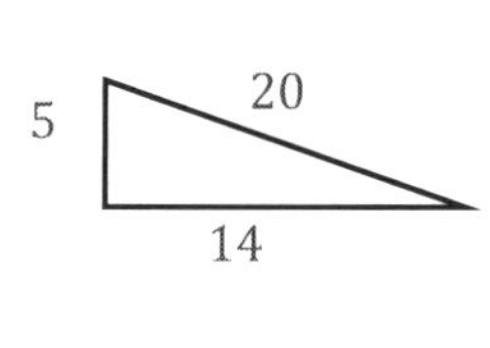

7) _____

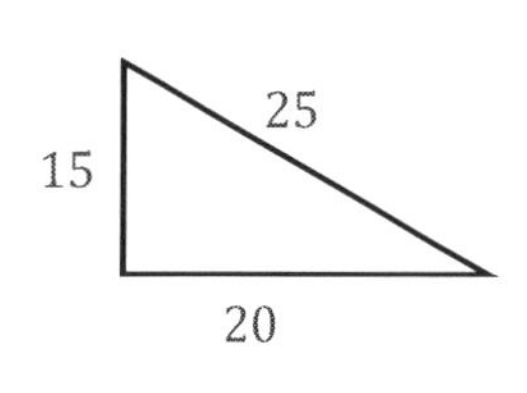

8) _____

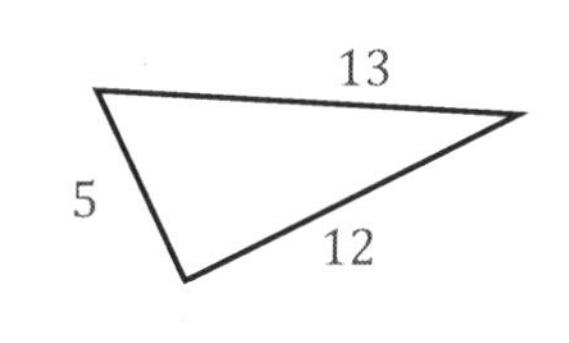

Find the missing side.

9) _____

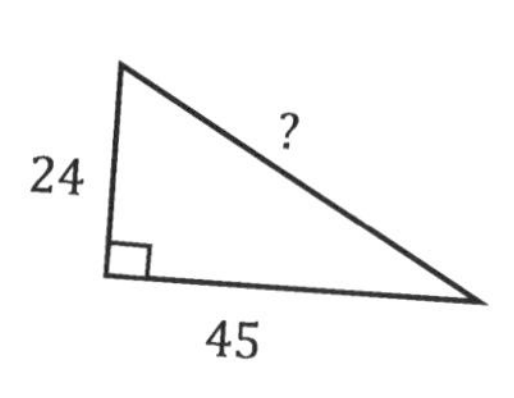

10) _____

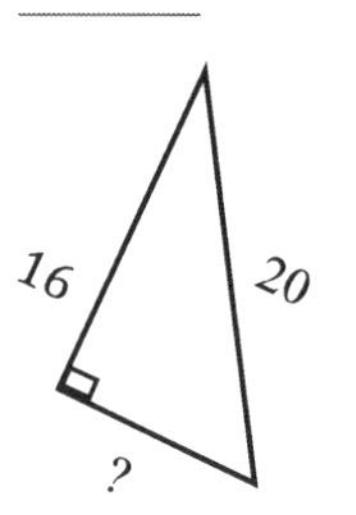

11) _____

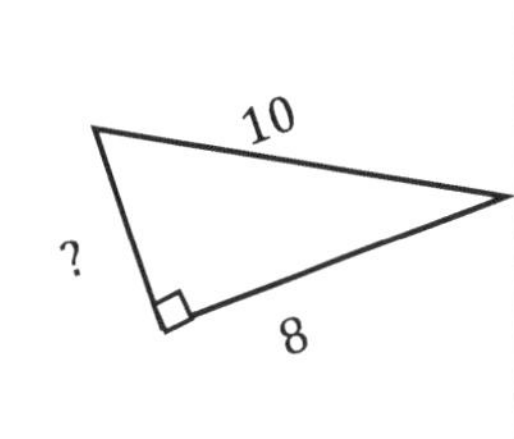

12) _____

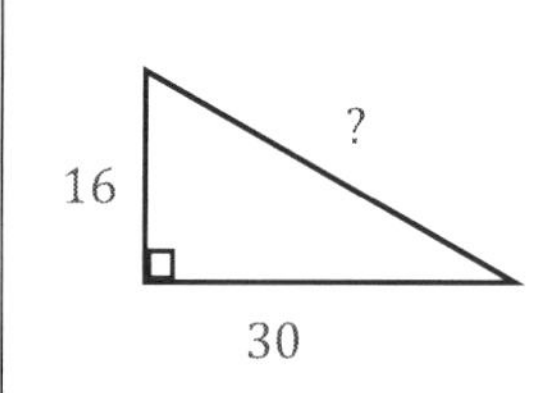

13) _____

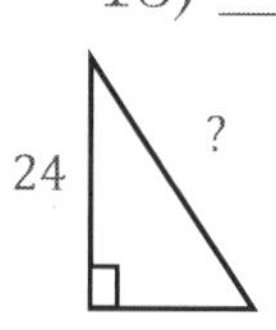

14) _____

15) _____

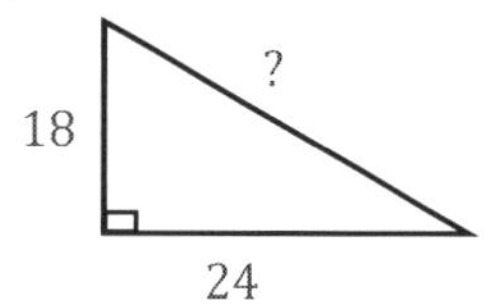

16) _____

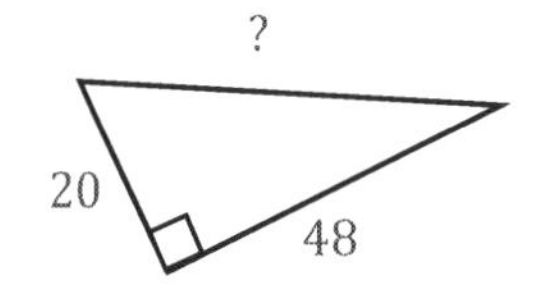

The Pythagorean Theorem - Answers

Do the following lengths form a right triangle?

1) yes

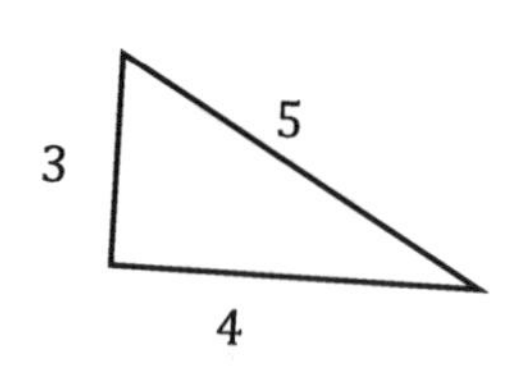

2) yes

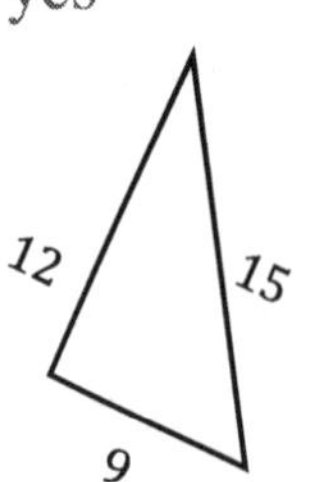

3) no

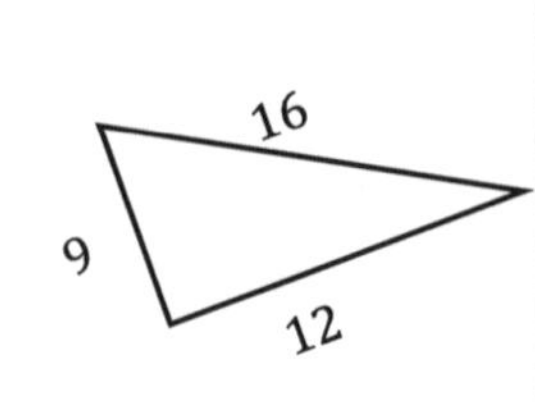

4) yes

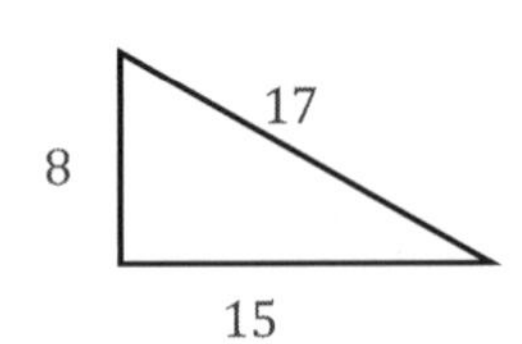

5) no

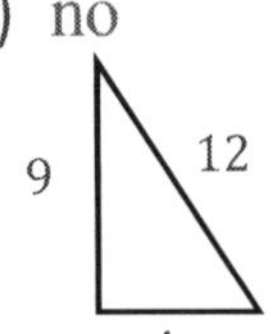

6) no

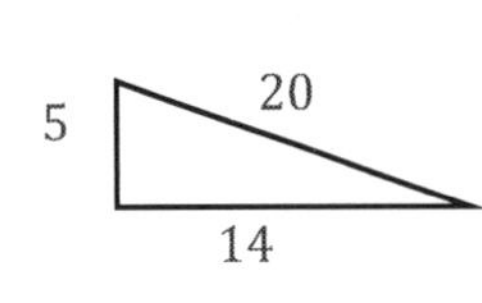

7) yes

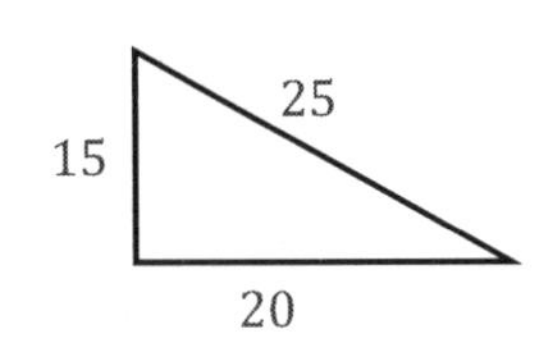

8) yes

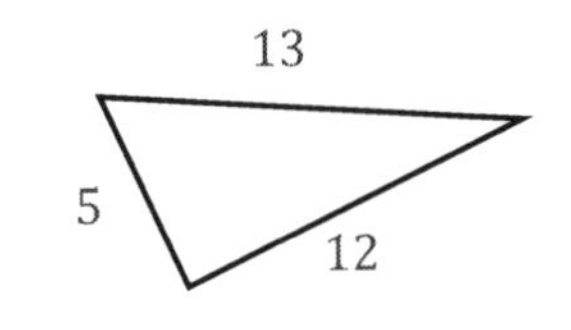

Find the missing side.

9) 51

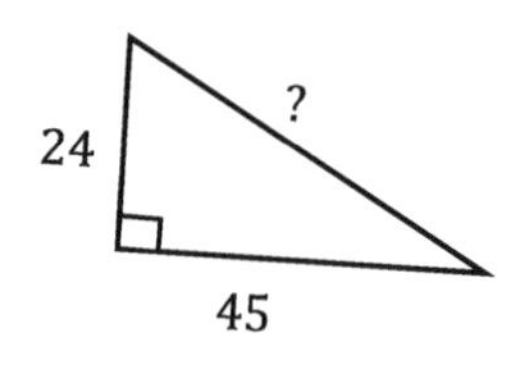

10) 12

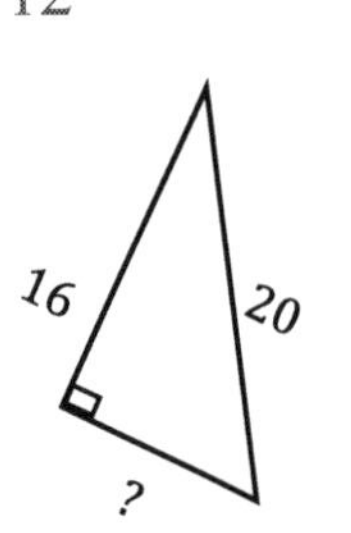

11) 6

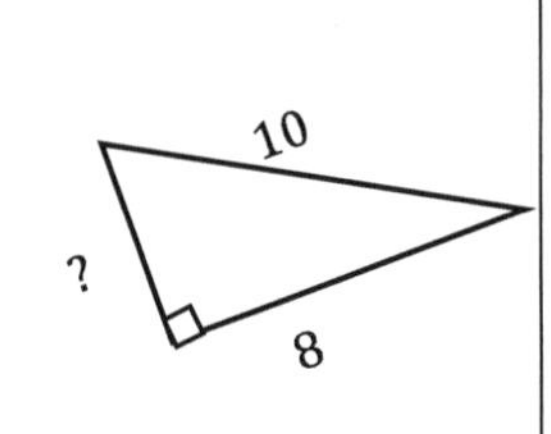

12) 34

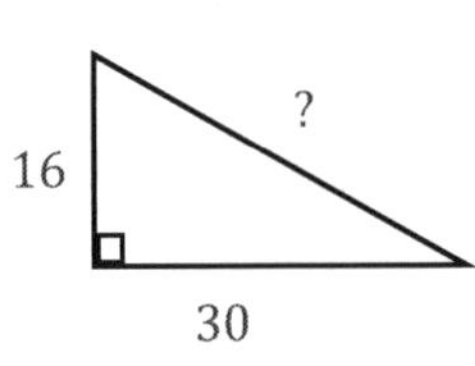

13) 26

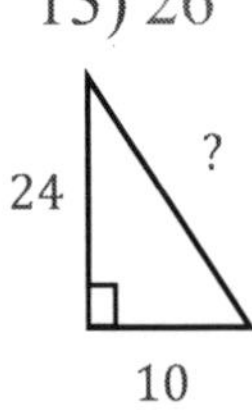

14) 13

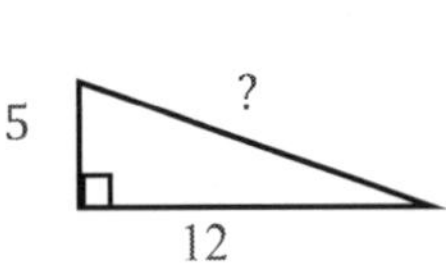

15) 30

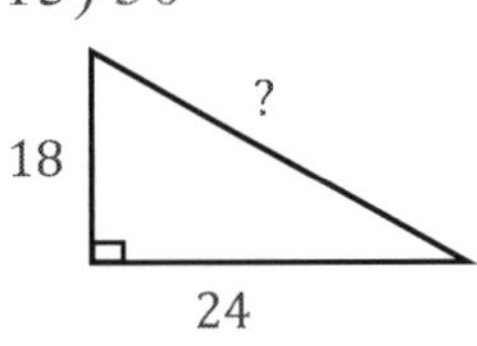

16)

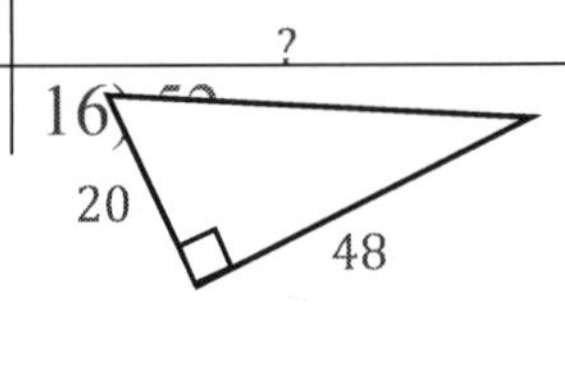

Triangles

Find the measure of the unknown angle in each triangle.

1) ______

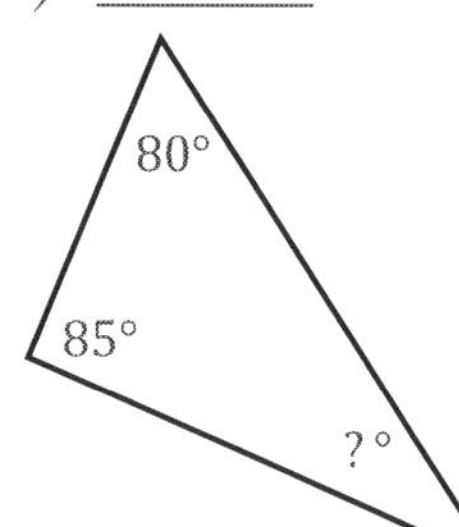

2) ______

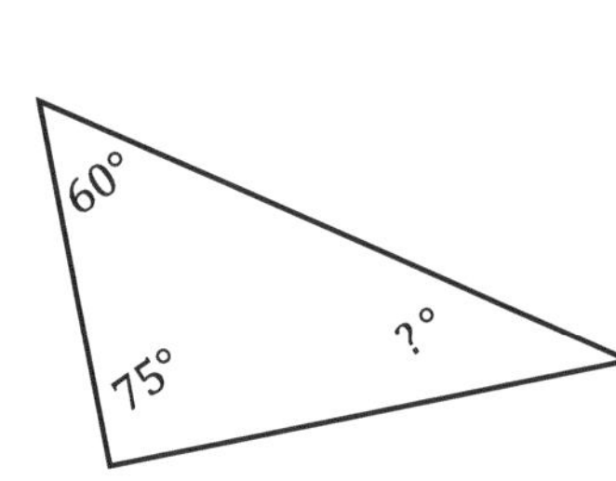

3) ______

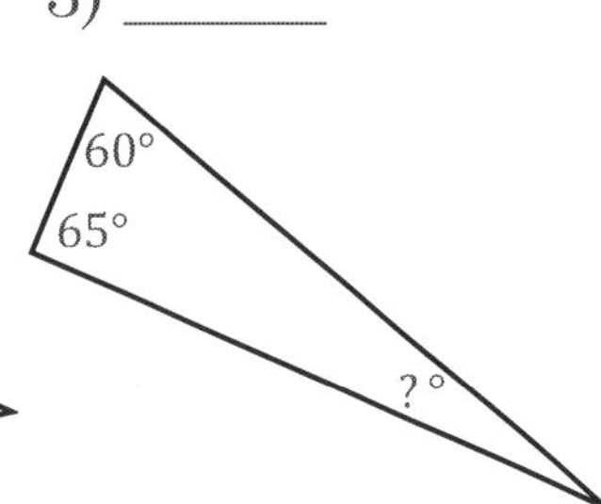

4) ______

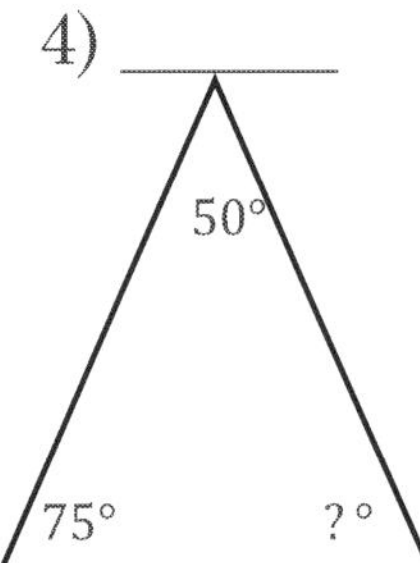

5) ______

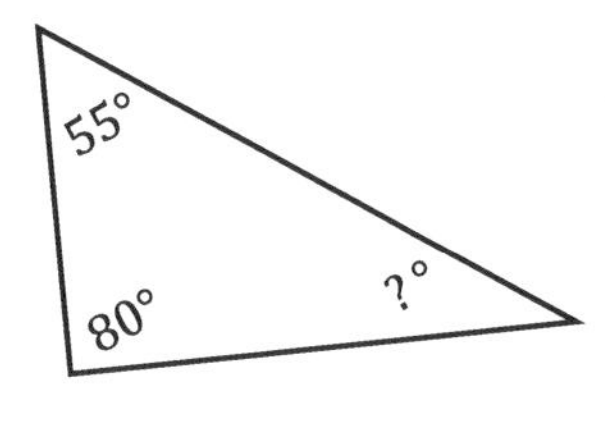

6) ______

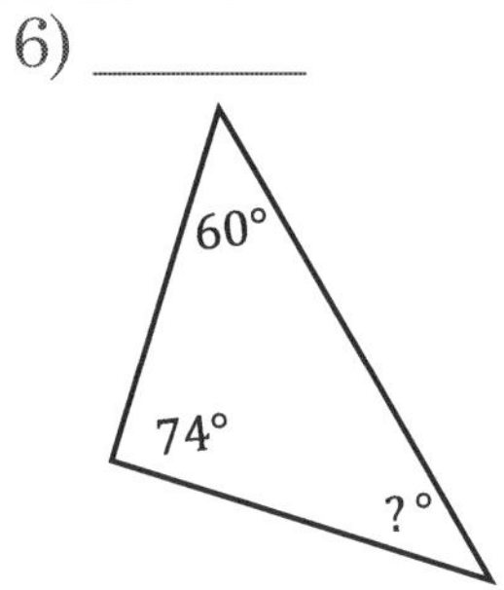

7) ______

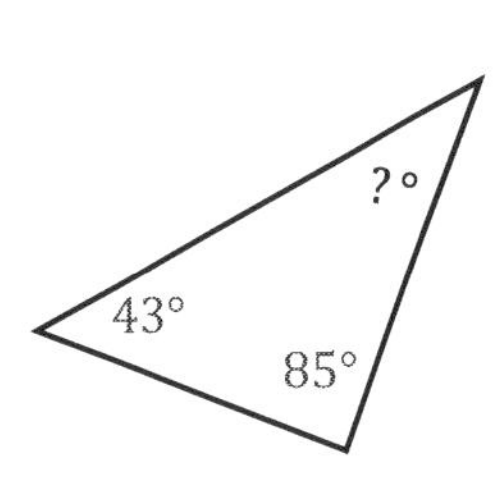

8) ______

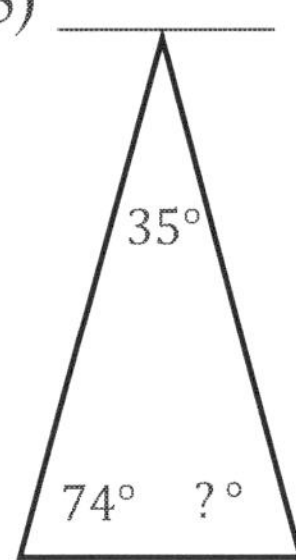

Find area of each triangle.

9) ______

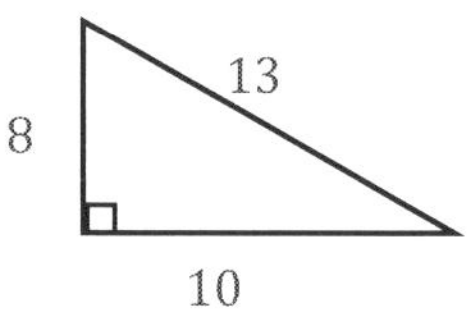

10) ______

14
15
8

11) ______

15 cm
8 cm
18 cm

12) ______

10
6 in
14 in

bit.ly/3haZrRg

Find more at

Triangles – Answers

Find the measure of the unknown angle in each triangle.

1) 15°

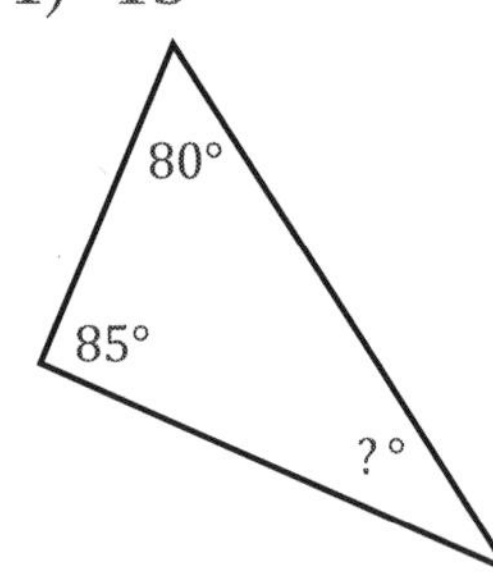

2) 45°

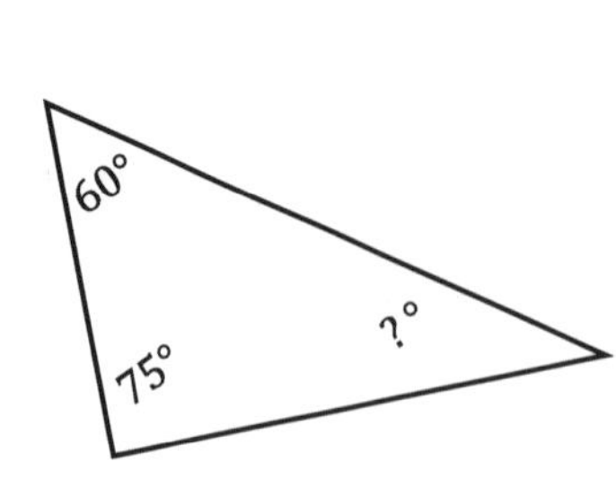

3) 55°

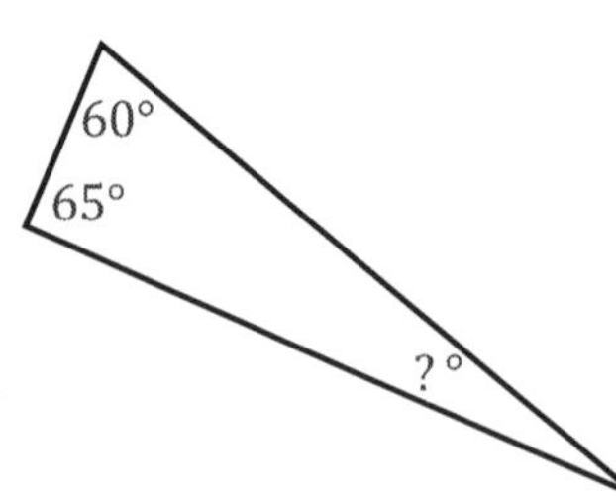

4) 55°

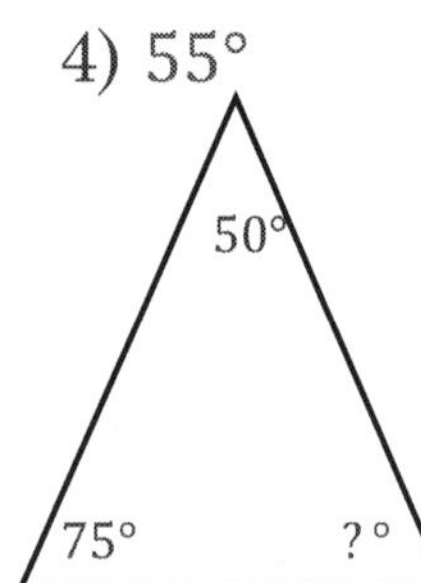

5) 45°

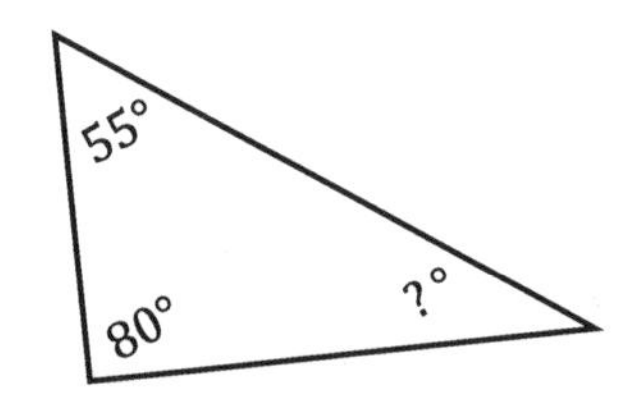

6) 46°

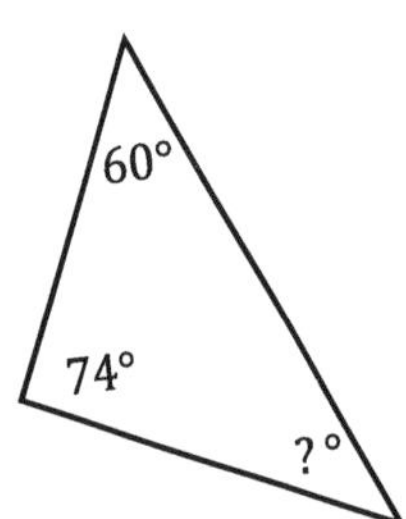

7) 52°

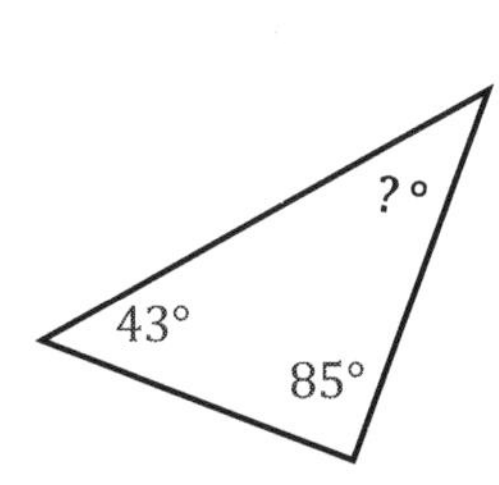

8) 71°

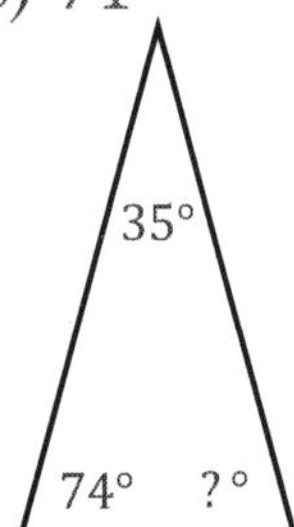

Find area of each triangle.

9) 40

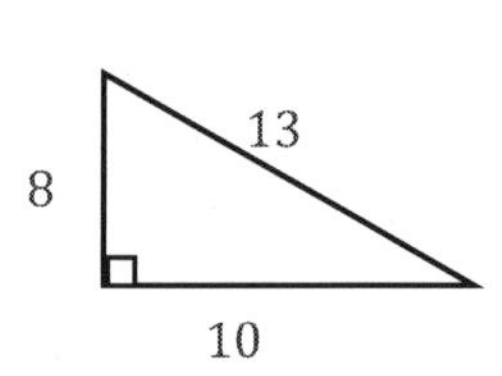

10) 56

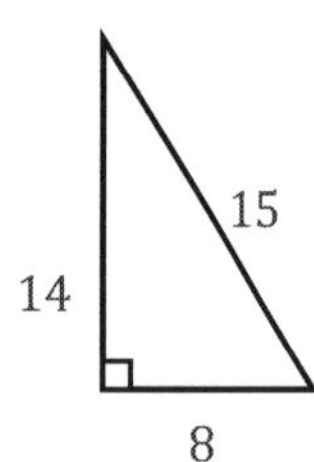

11) $72\ cm^2$

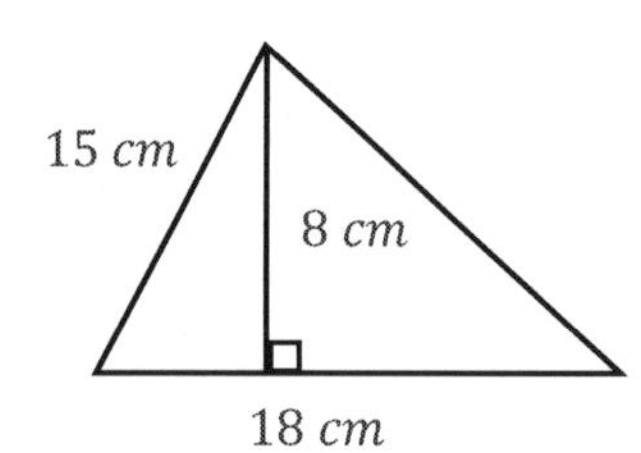

12) $42\ in^2$

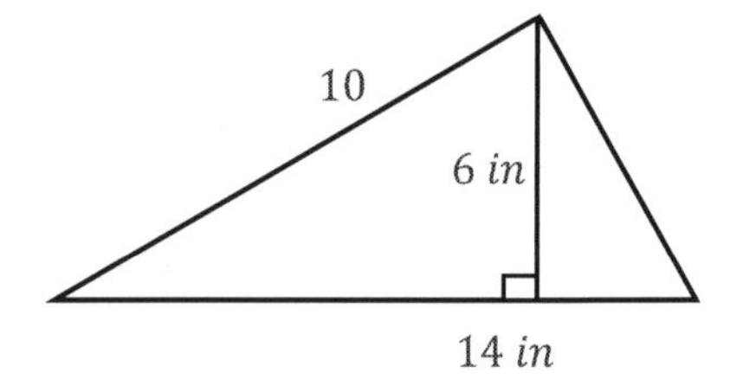

Polygons

Find the perimeter of each shape.

1) (square) ______

5 cm

2) ______

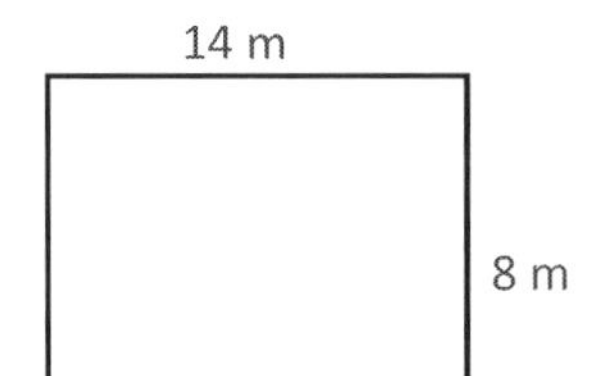

3) ______

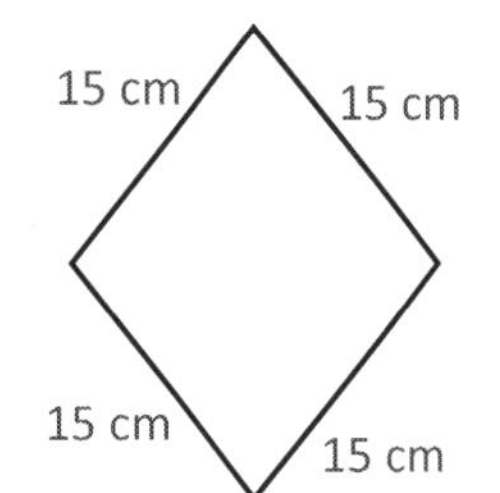

4) (square) ______

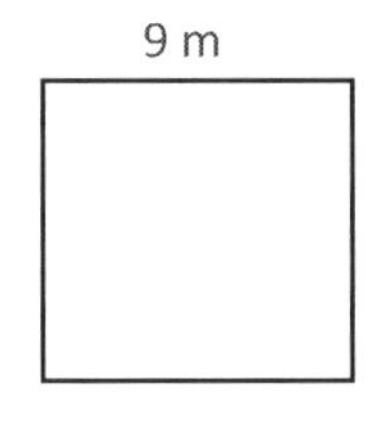

5) *(regular hexagon)* ______

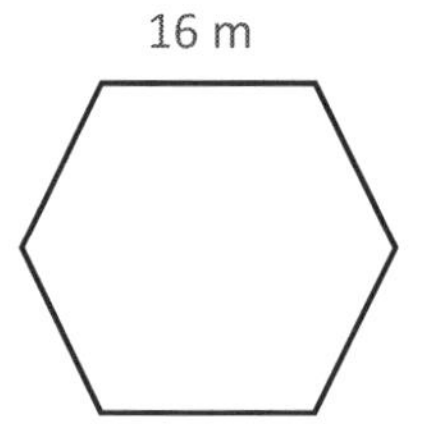

6) ______

14 m
12 m
12 m
18 m

7) *(parallelogram)* ______

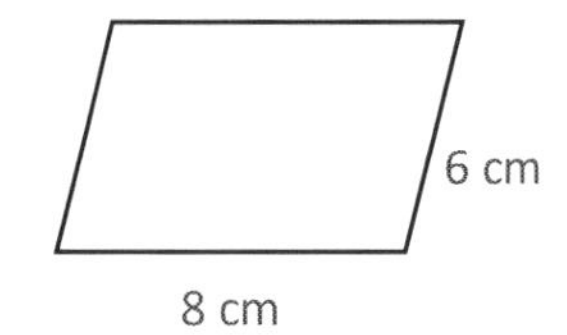

8) *(regular hexagon)* ______

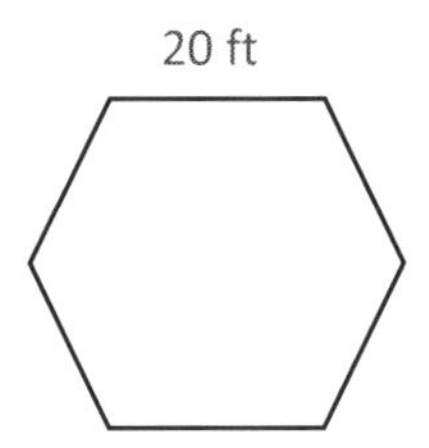

9) ______

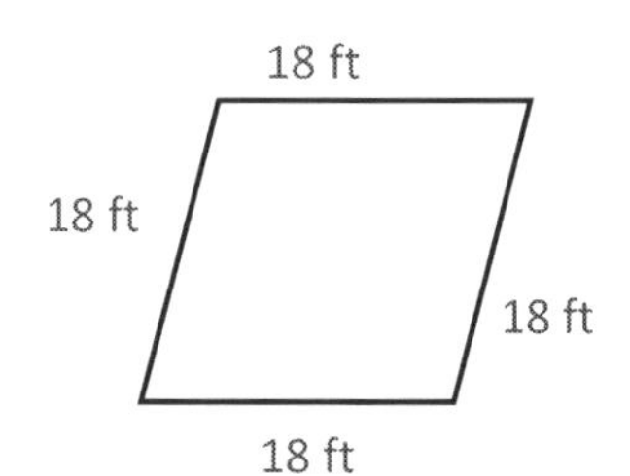

10) ______

20 in
16 in

11) ______

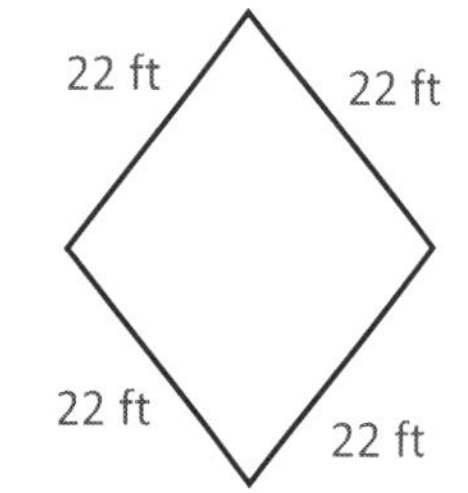

12) *(regular hexagon)* ______

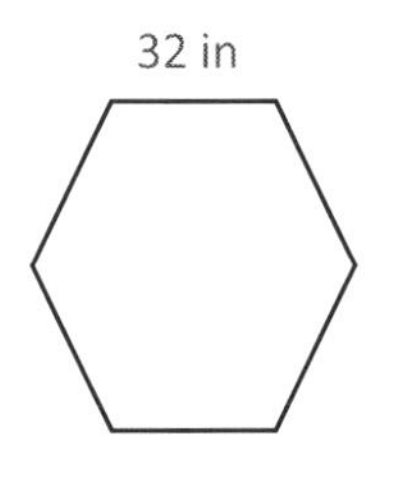

Polygons - Answers

Find the perimeter of each shape.

1) (square) $20\ cm$

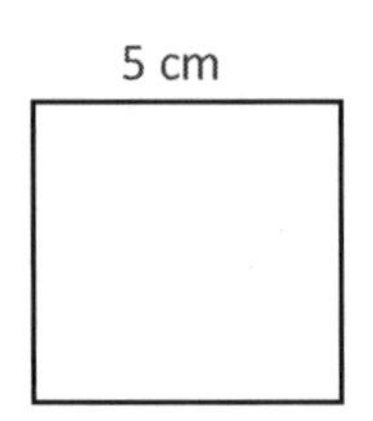

2) $44\ m$

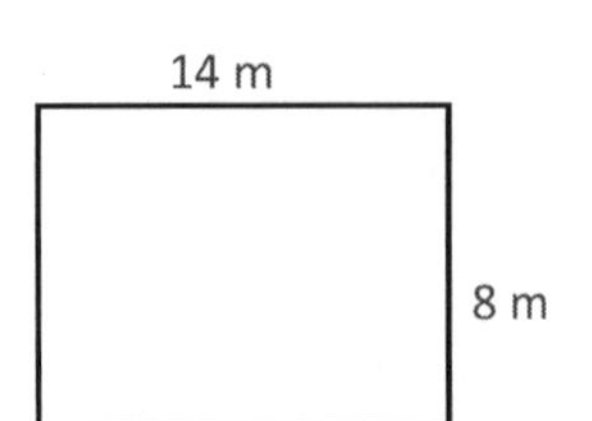

3) $60\ cm$

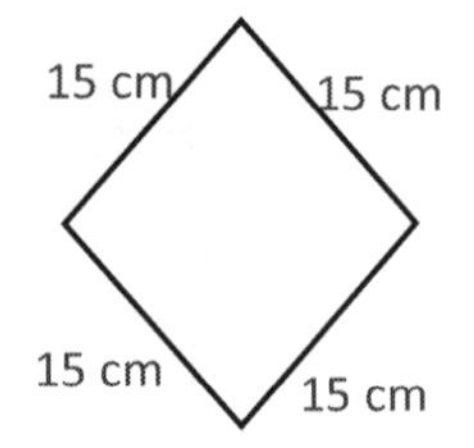

4) (square) $36\ m$

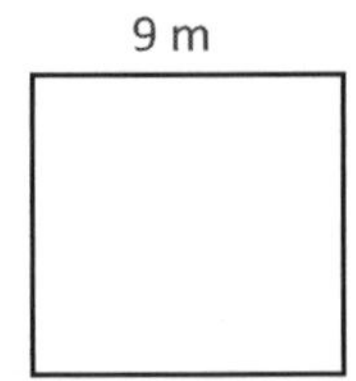

5) *(regular hexagon)* $96\ m$

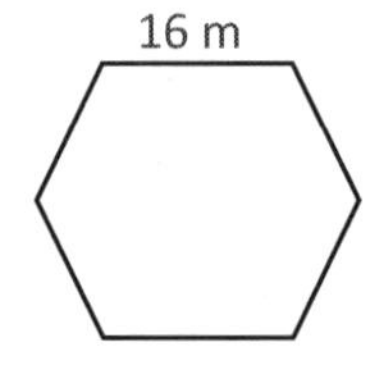

6) $56\ m$

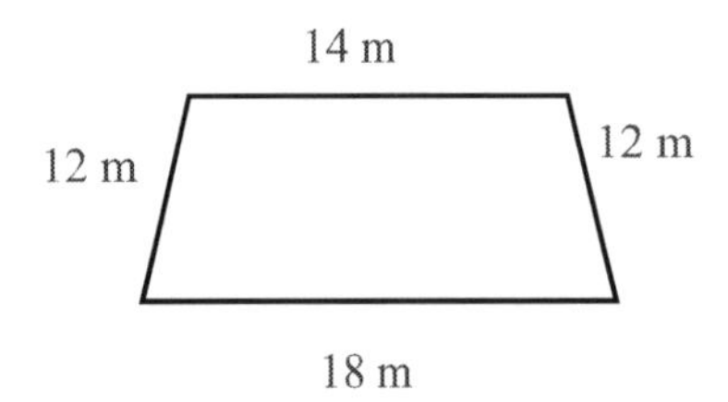

7) *(parallelogram)* $28\ cm$

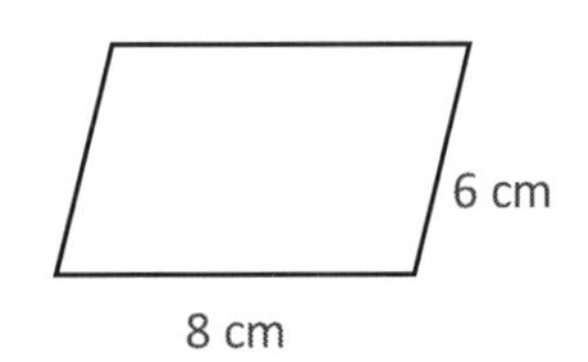

8) *(regular hexagon)* $120\ ft$

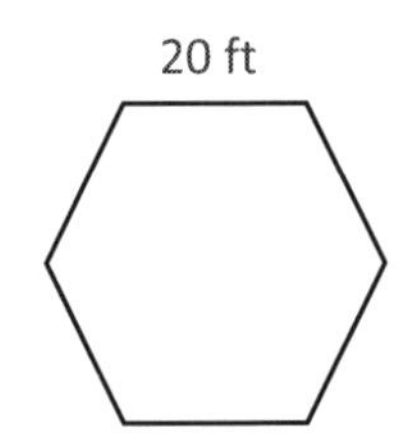

9) $72\ ft$

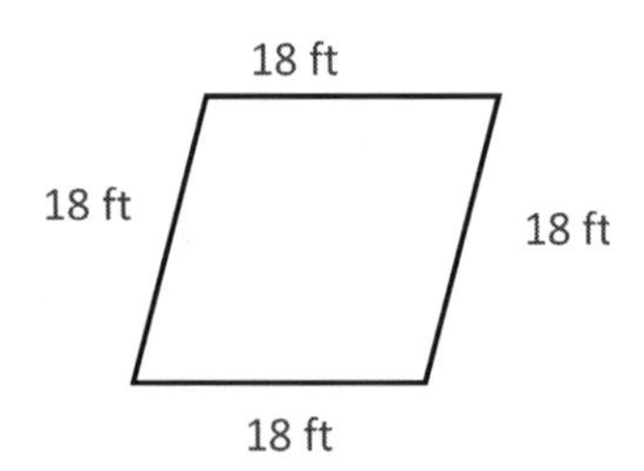

10) $72\ in$

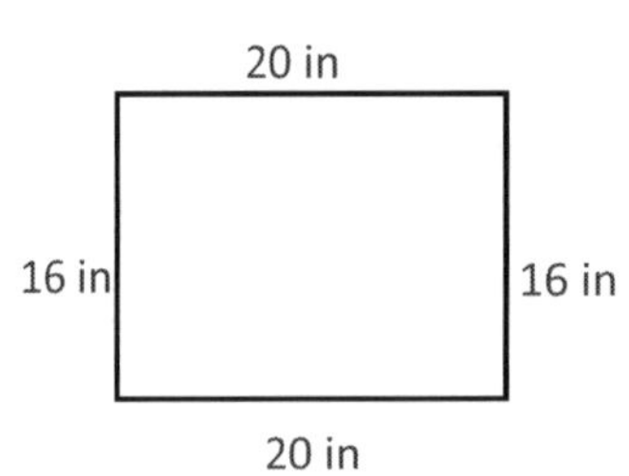

11) $88\ ft$

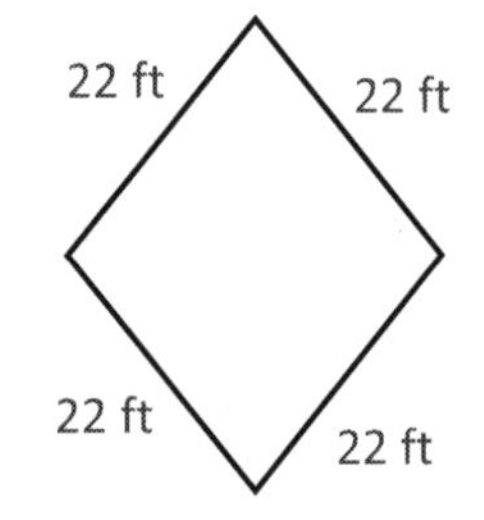

12) *(regular hexagon)* $192\ in$

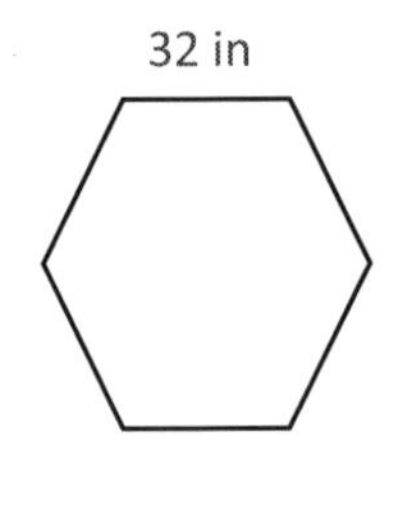

Circles

Find the Circumference of each circle. (π = 3.14)

1) ____ 2) ____ 3) ____ 4) ____ 5) ____ 6) ____

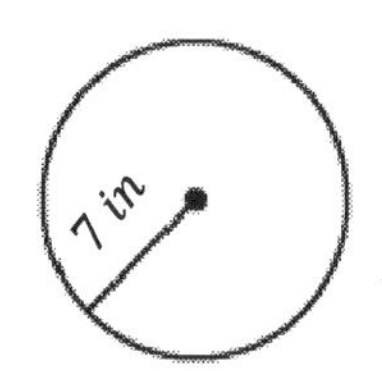

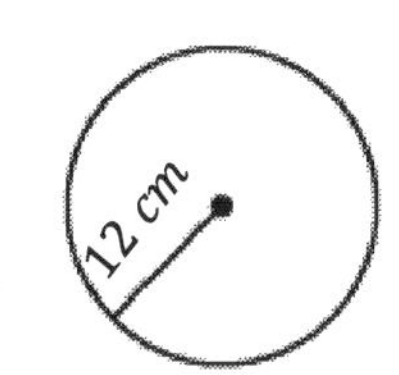

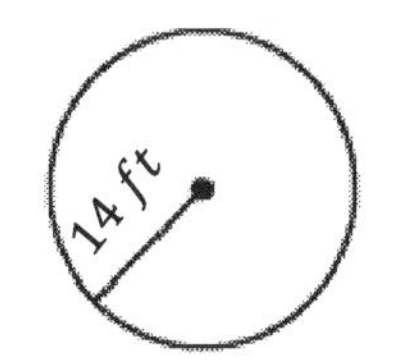

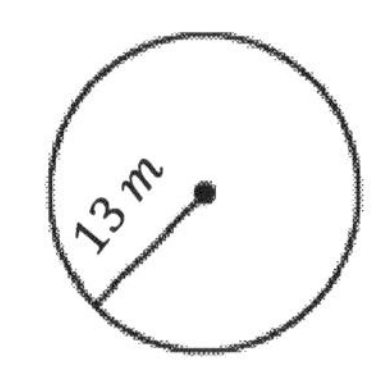

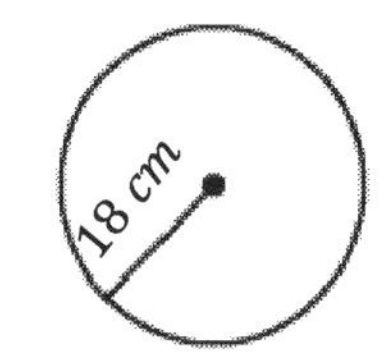

7) ____ 8) ____ 9) ____ 10) ____ 11) ____ 12) ____

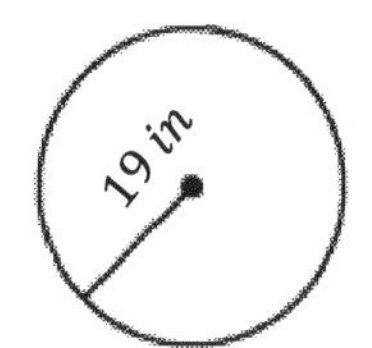

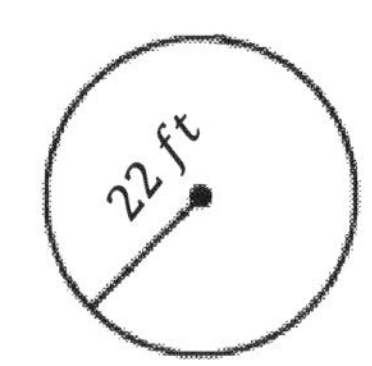

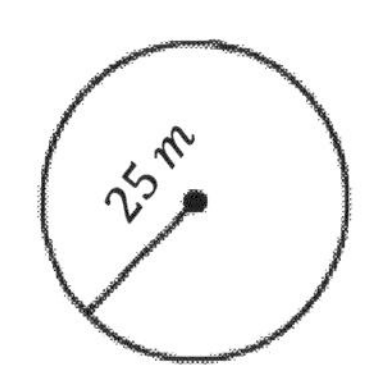

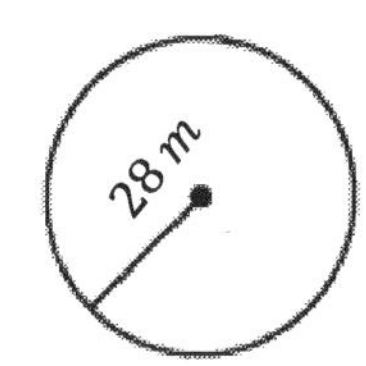

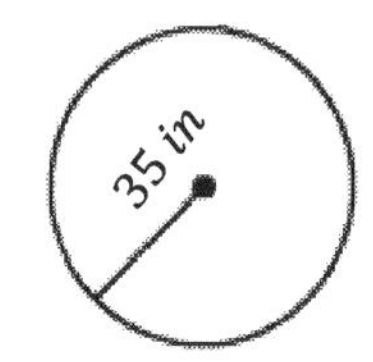

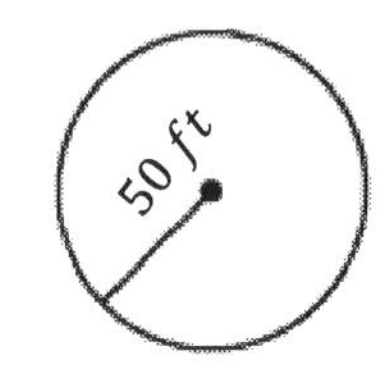

 ***Complete the table below.*** (π = 3.14)

	Radius	Diameter	Circumference	Area
Circle 1	2 *inches*	4 *inches*	12.56 *inches*	12.56 *square inches*
Circle 2		8 *meters*		
Circle 3				113.04 *square ft*
Circle 4			50.24 *miles*	
Circle 5		9 *km*		
Circle 6	7 *cm*			
Circle 7		10 *feet*		
Circle 8				615.44 *square meters*
Circle 9			81.64 *inches*	
Circle 10	12 *feet*			

Circles - Answers

Find the Circumference of each circle. (π = 3.14)

1) 43.96 *in*	2) 75.36 *cm*	3) 87.92 *ft*	4) 81.64 *m*	5) 113.04 *cm*	6) 94.2 *miles*
7 in	12 cm	14 ft	13 m	18 cm	15 miles
7) 119.32 *in*	8) 138.16 *ft*	9) 157 *m*	10) 175.84 *m*	11) 219.8 *in*	12) 314 *ft*
19 in	22 ft	25 m	28 m	35 in	50 ft

Complete the table below. (π = 3.14)

	Radius	Diameter	Circumference	Area
Circle 1	2 *inches*	4 *inches*	12.56 *inches*	12.56 *square inches*
Circle 2	4 *meters*	8 *meters*	25.12 *meters*	50.24 *square meters*
Circle 3	6 ft	12 *ft*	37.68	113.04 *square ft*
Circle 4	8 *miles*	16 *miles*	50.24 *miles*	200.96 *square miles*
Circle 5	4.5 *km*	9 *km*	28.26 *km*	63.585 *square km*
Circle 6	7 *cm*	14 *cm*	43.96 *cm*	153.86 *square cm*
Circle 7	5 *feet*	10 *feet*	31.4 *feet*	78.5 *square feet*
Circle 8	14 *m*	28 *m*	87.92 *m*	615.44 *square meters*
Circle 9	13 *in*	26 *in*	81.64 *inches*	530.66 *square inches*
Circle 10	12 *feet*	24 *feet*	75.36 *feet*	452.16 *square feet*

Cubes

Find the volume of each cube.

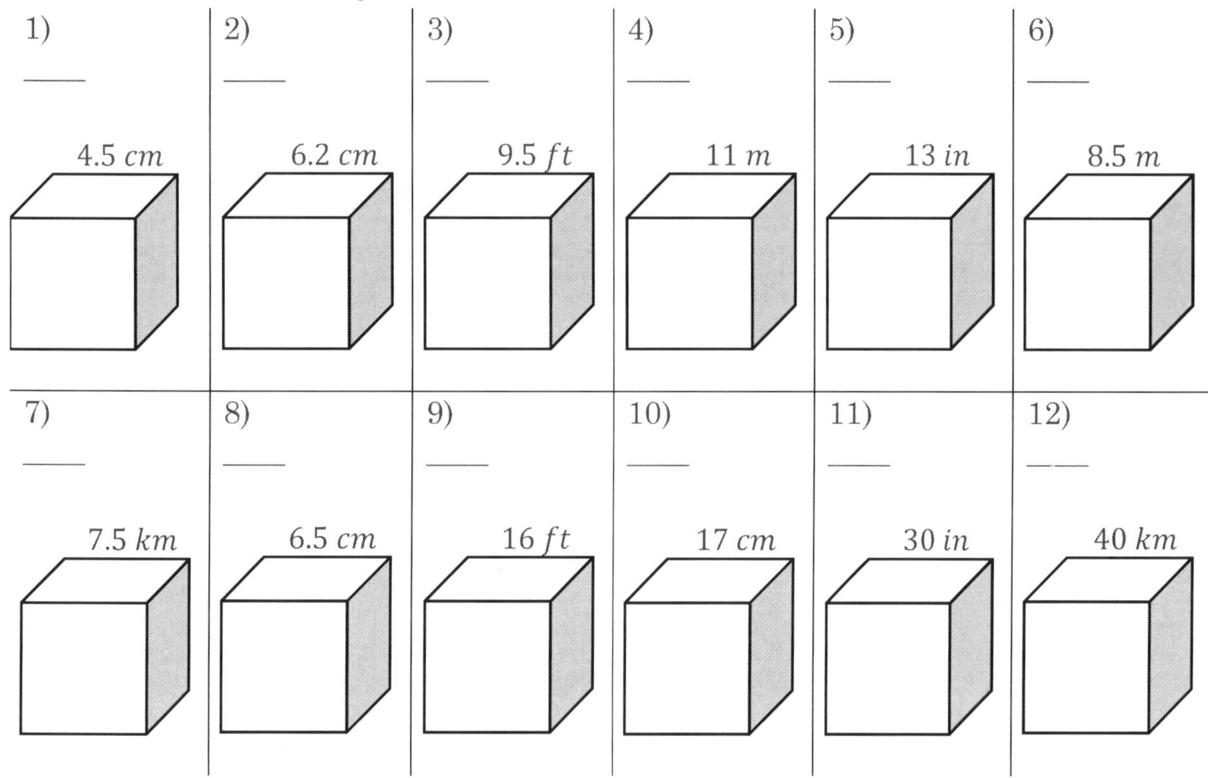

Find the surface area of each cube.

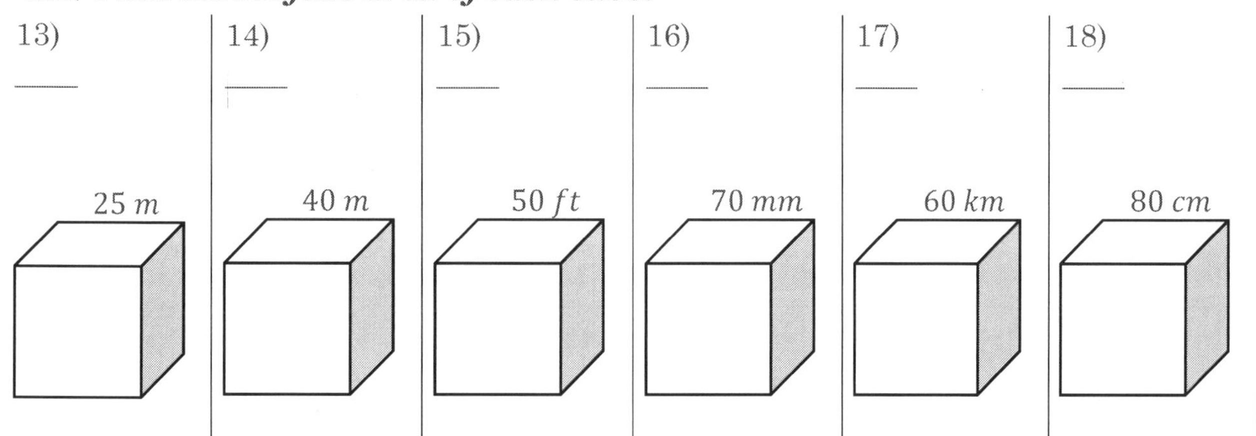

Cubes - Answers

Find the volume of each cube.

1) $91.125\ cm^3$ — 4.5 cm

2) $238.328\ cm^3$ — 6.2 cm

3) $857.375\ ft^3$ — 9.5 ft

4) $1{,}331\ m^3$ — 11 m

5) $2{,}197\ in^3$ — 13 in

6) $614.125\ m^3$ — 8.5 m

7) $421.875\ km^3$ — 7.5 km

8) $274.625\ cm^3$ — 6.5 cm

9) $4{,}096\ ft^3$ — 16 ft

10) $4{,}913\ cm^3$ — 17 cm

11) $27{,}000\ in^3$ — 30 in

12) $64{,}000\ km^3$ — 40 km

Find the surface area of each cube.

13) $3{,}750\ m^2$

14) $9{,}600\ m^2$

15) $15{,}000\ ft^2$

16) $29{,}400\ mm^2$

17) $21{,}600\ km^2$

18) $38{,}400\ cm^2$

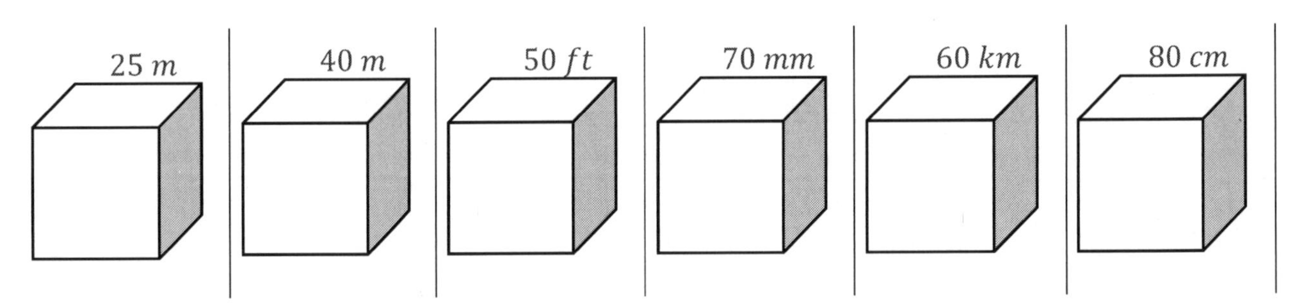

Trapezoids

 ***Find the area of each trapezoid.***

1) _____ 2) _____ 3) _____ 4) _____

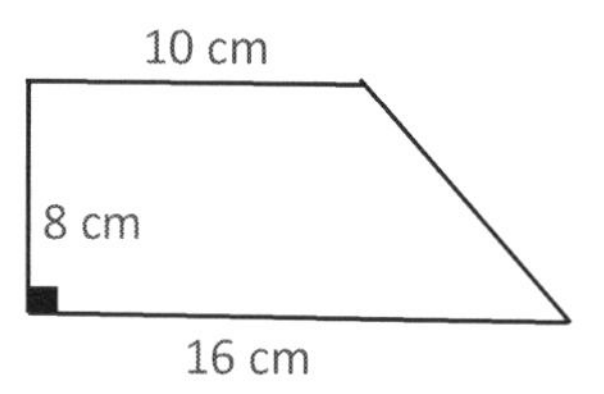

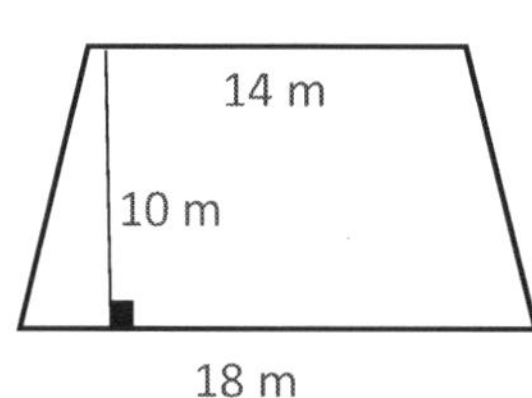

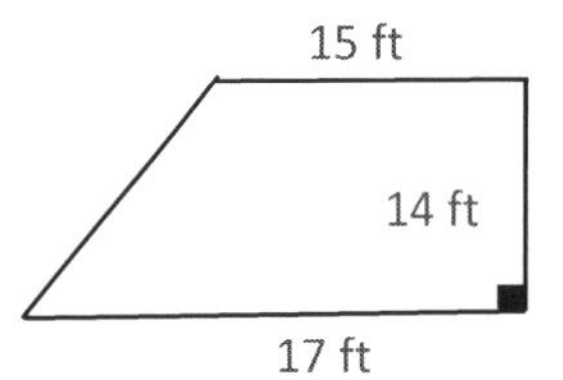

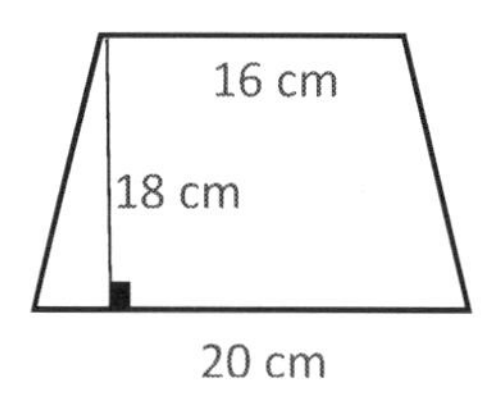

5) _____ 6) _____ 7) _____ 8) _____

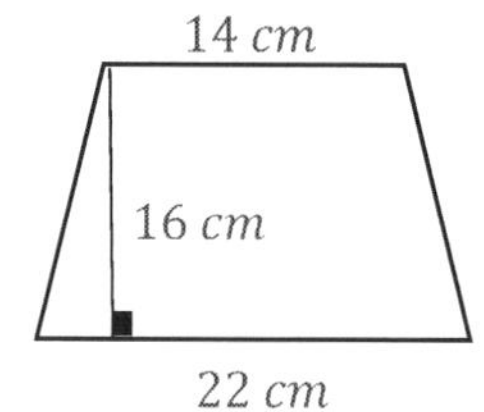

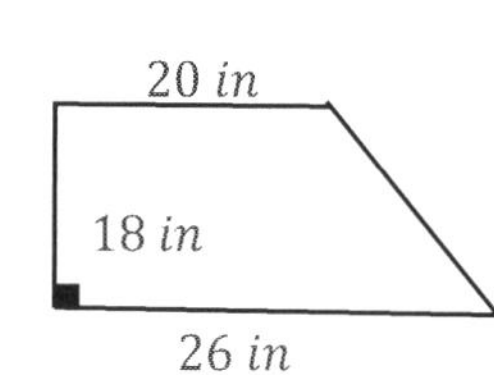

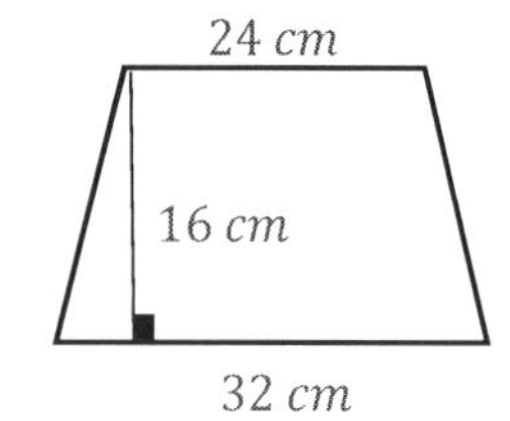

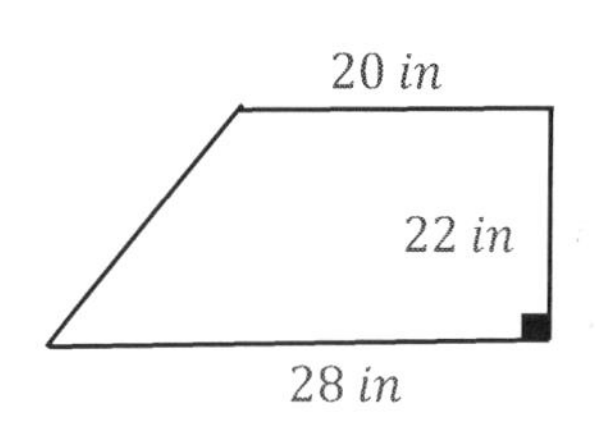

 ***Solve.***

9) A trapezoid has an area of 80 $cm2$ and its height is 8 cm and one base is 12 cm. What is the other base length? ________________

10) If a trapezoid has an area of 120 $ft2$ and the lengths of the bases are 14 ft and 16 ft, find the height. ________________

11) If a trapezoid has an area of 160 $m2$ and its height is 10 m and one base is 14 m, find the other base length. ________________

12) The area of a trapezoid is 504 $ft2$ and its height is 24 ft. If one base of the trapezoid is 14 ft, what is the other base length?

____

Trapezoids - Answers

 ***Find the area of each trapezoid.***

1) $\mathbf{104\ cm^2}$ 2) $\mathbf{160\ m^2}$ 3) $\mathbf{224\ ft^2}$ 4) $\mathbf{324\ cm^2}$

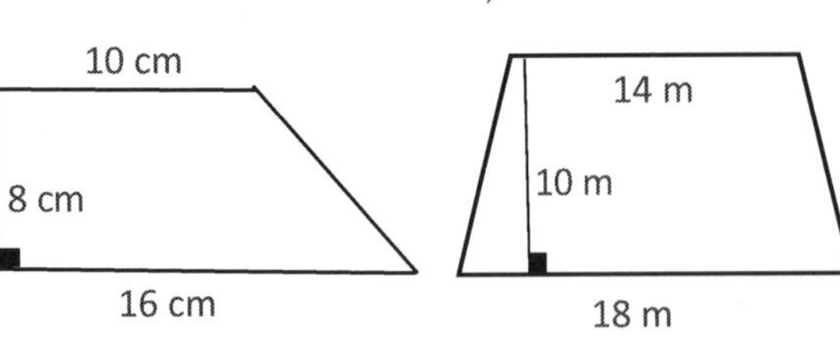

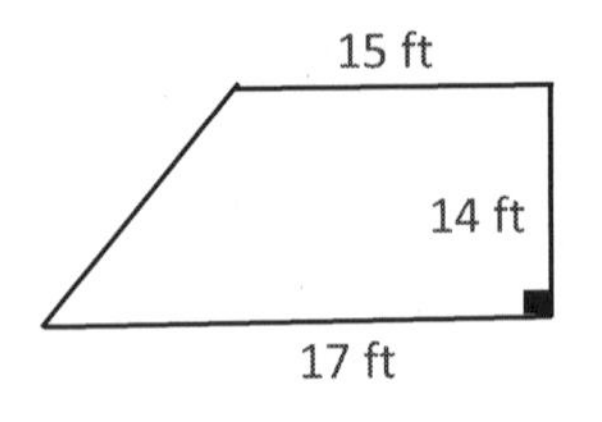

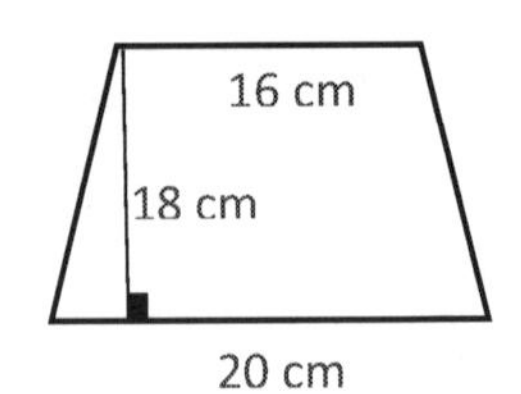

5) $\mathbf{288\ cm^2}$ 6) $414\ in^2$ 7) $448\ cm^2$ 8) $528\ in^2$

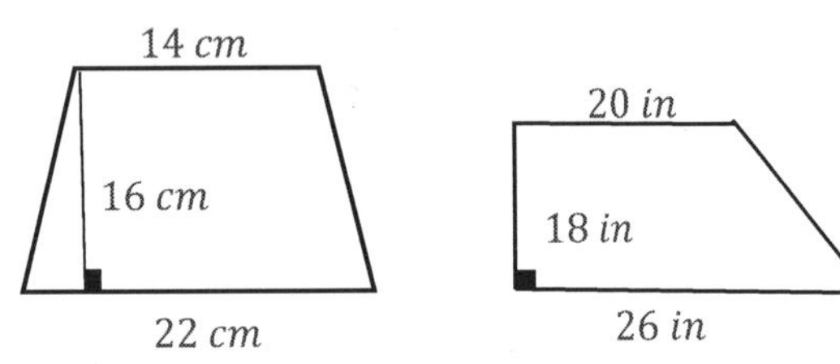

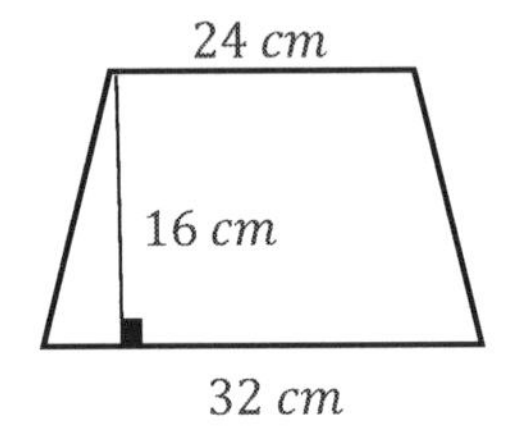

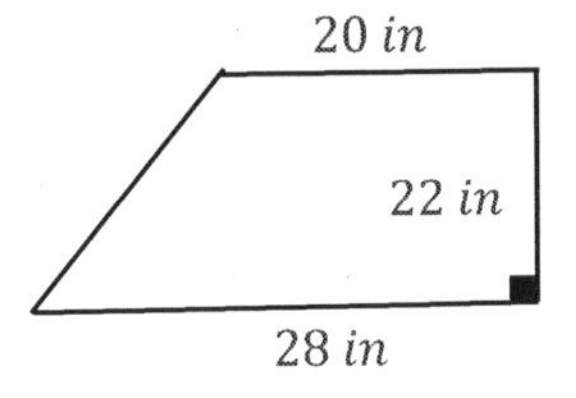

 ***Solve.***

9) A trapezoid has an area of 80 $cm2$ and its height is 8 cm and one base is 12 cm. What is the other base length? 8 cm

10) If a trapezoid has an area of 120 $ft2$ and the lengths of the bases are14 ft and 16ft, find the height. 8 ft

11) If a trapezoid has an area of 160 $m2$ and its height is 10 m and one base is 14 m, find the other base length. 18 m

12) The area of a trapezoid is 504 $ft2$ and its height is 24 ft. If one base of the trapezoid is 14 ft, what is the other base length? 28 ft

Rectangular Prisms

✍ ***Find the volume of each Rectangular Prism.***

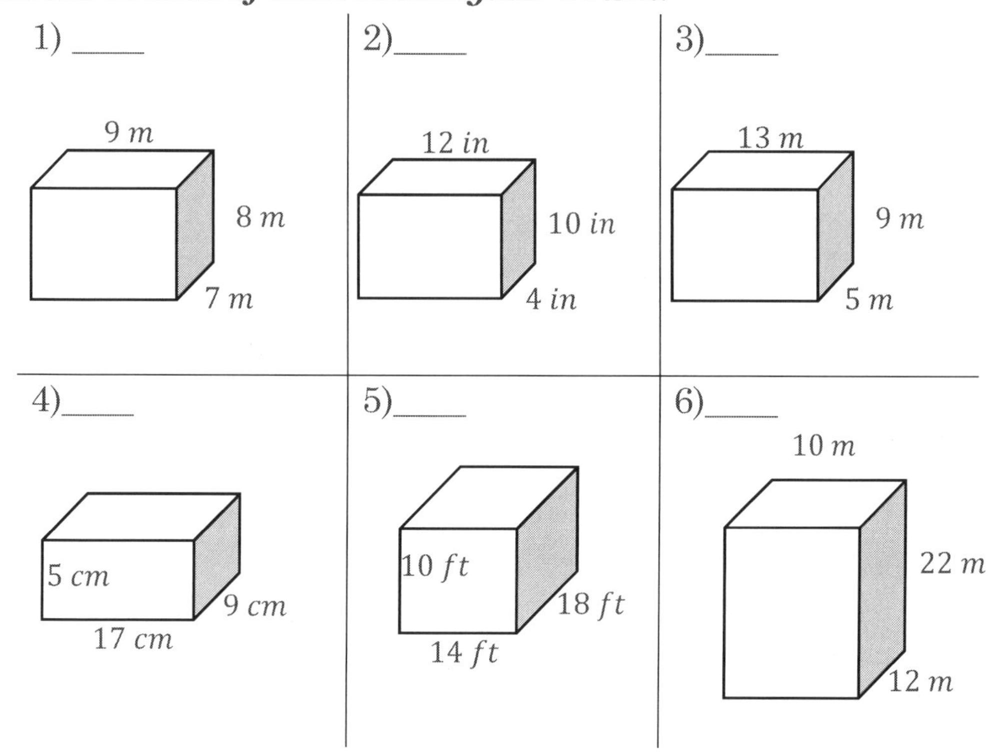

✍ ***Find the surface area of each Rectangular Prism.***

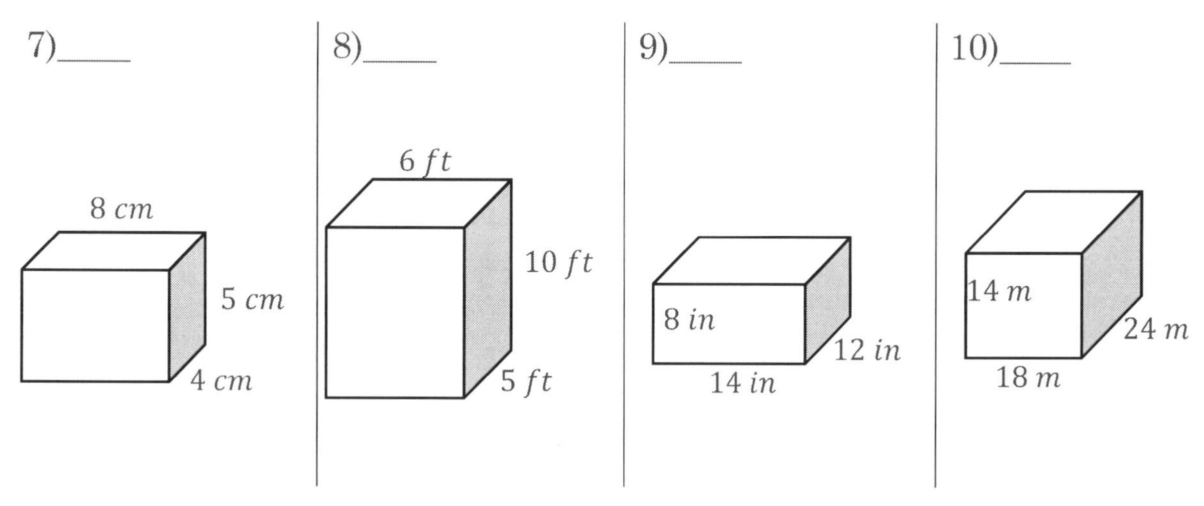

Rectangular Prisms - Answers

Find the volume of each Rectangular Prism.

1) $504\ m^3$ 2) $480\ in^3$ 3) $585\ m^3$

4) $765\ cm^3$ 5) $2{,}520\ ft^3$ 6) $2{,}640\ m^3$

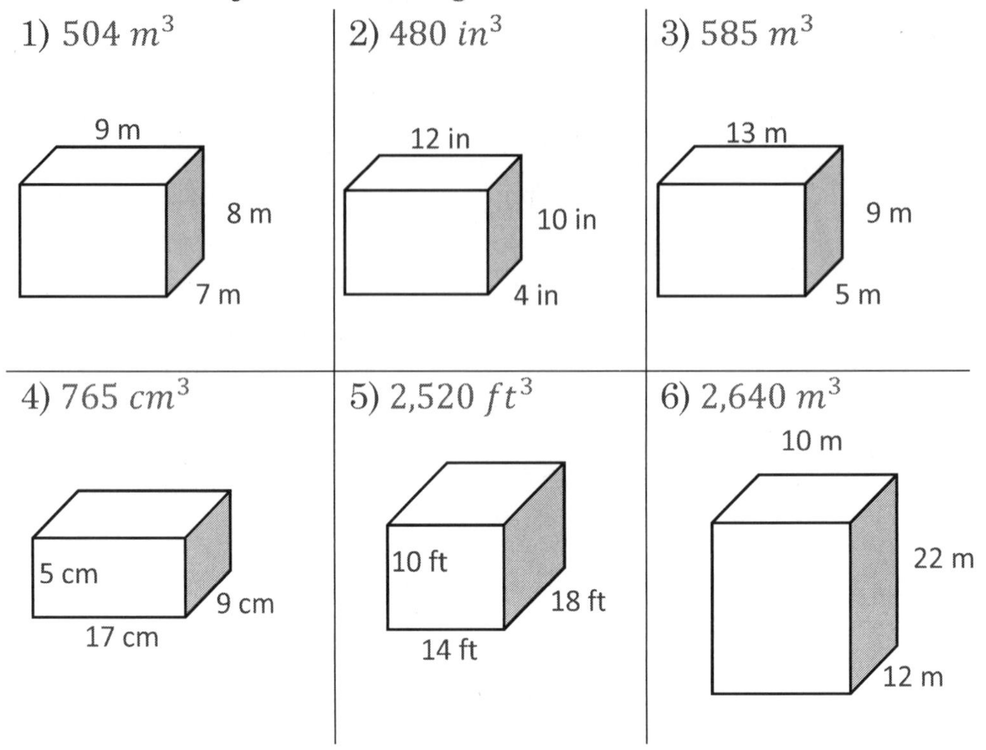

Find the surface area of each Rectangular Prism.

7) $184\ cm^2$ 8) $280\ ft^2$ 9) $752\ in^2$ 10) $2{,}040\ m^2$

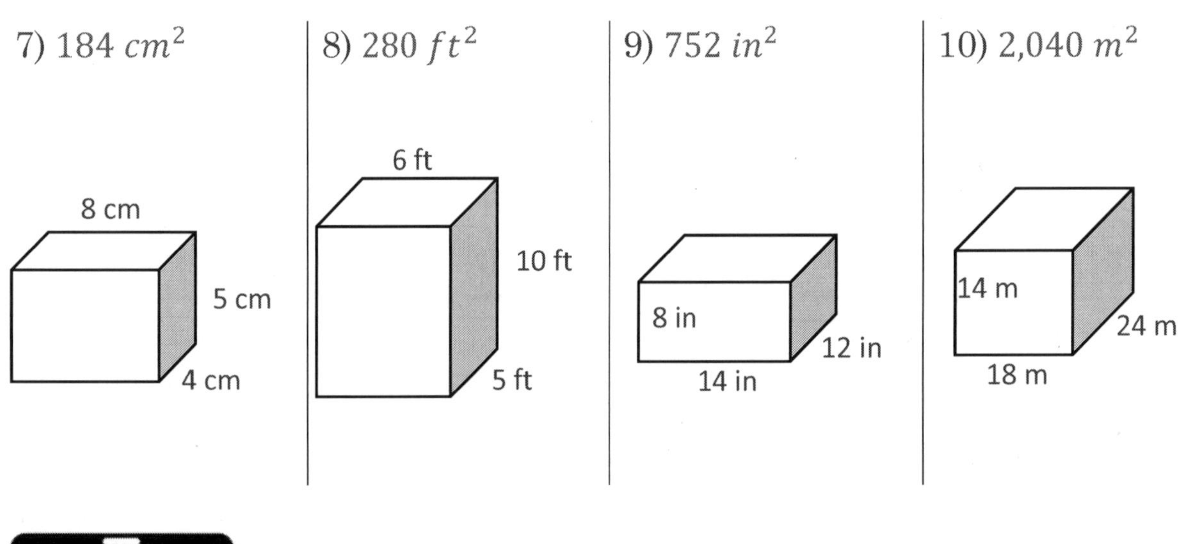

Cylinder

Find the volume of each Cylinder. ($\pi = 3.14$)

1) _____

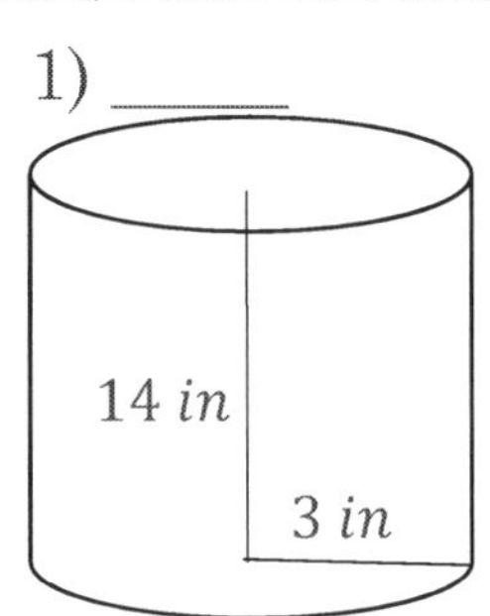

2) _____

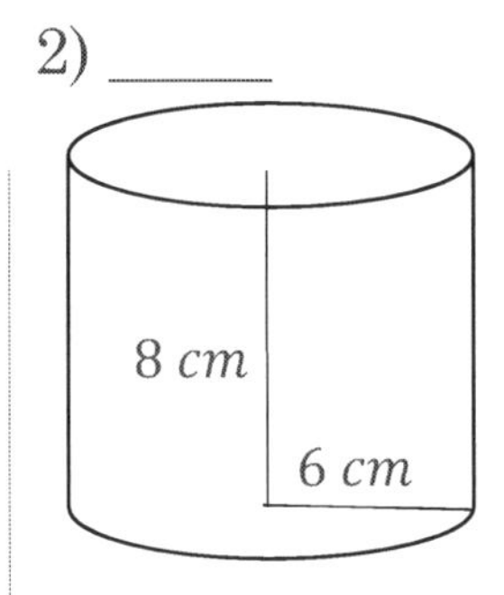

3) _____

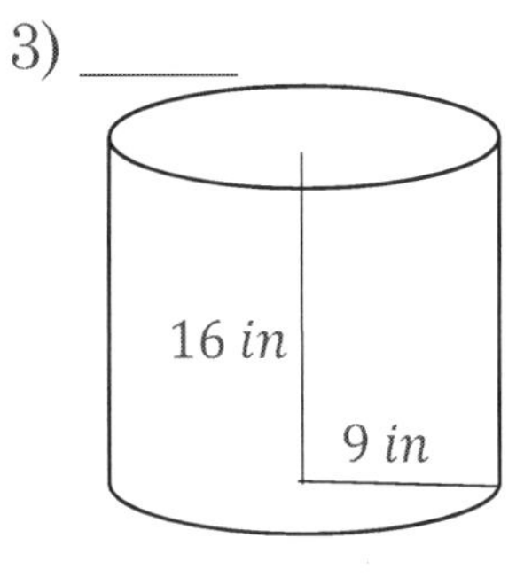

4) _____

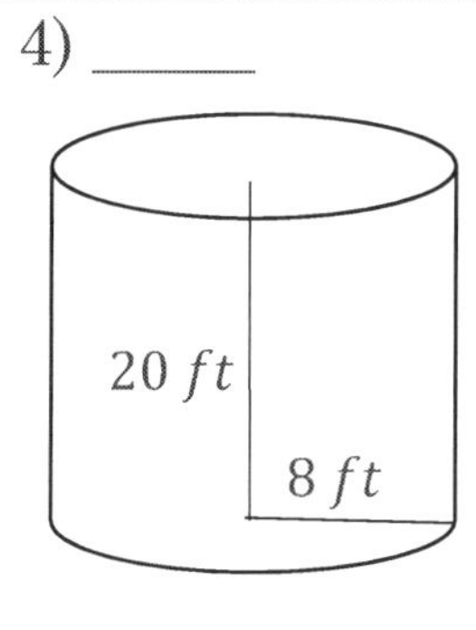

5) _____

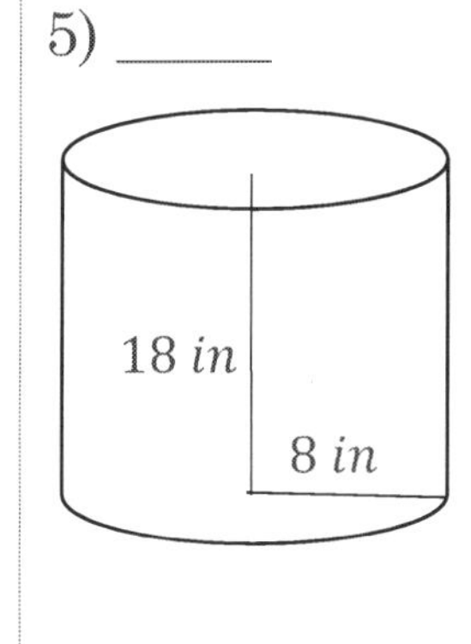

6) _____

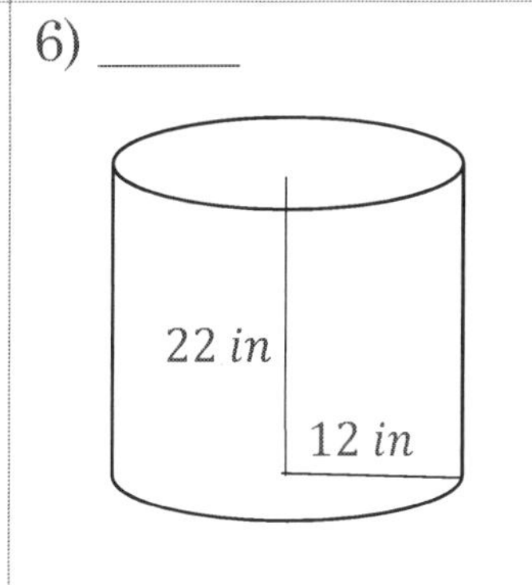

Find the surface area of each Cylinder. ($\pi = 3.14$)

7) _____

10 in
5 in

8) _____

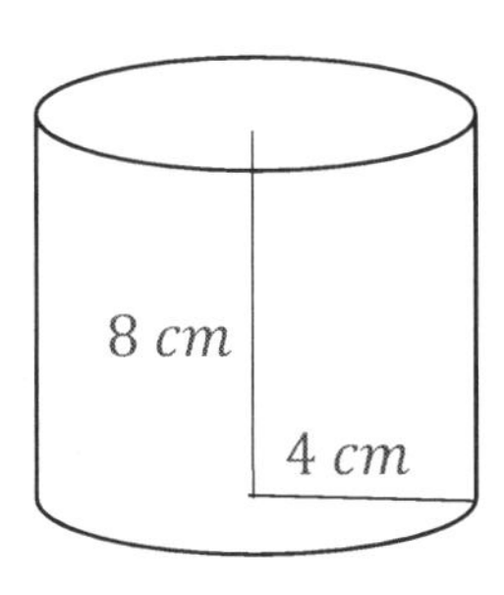

9) _____

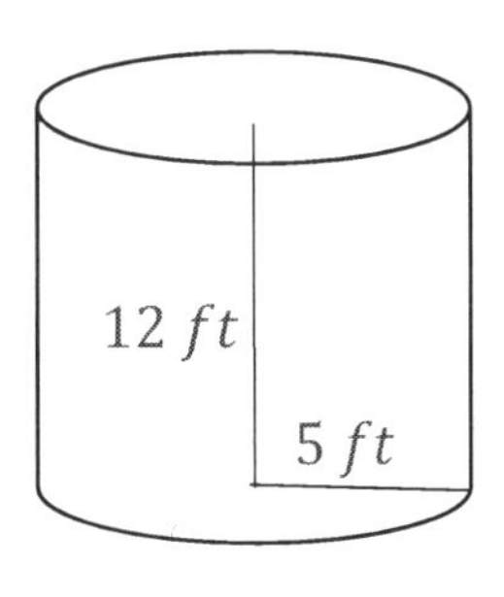

10)_____

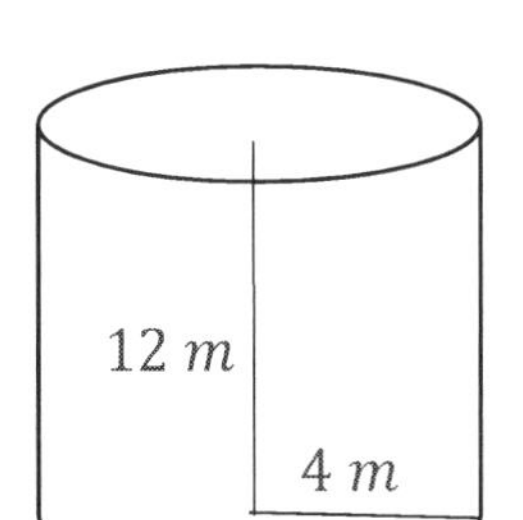

Cylinder - Answers

Find the volume of each Cylinder. (π = 3.14)

1) $\mathbf{395.64}\ in^3$

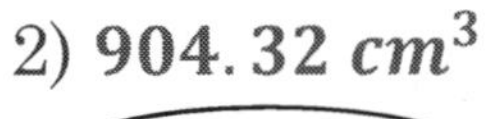

2) $\mathbf{904.32}\ cm^3$

3) $\mathbf{4,069.44}\ in^3$

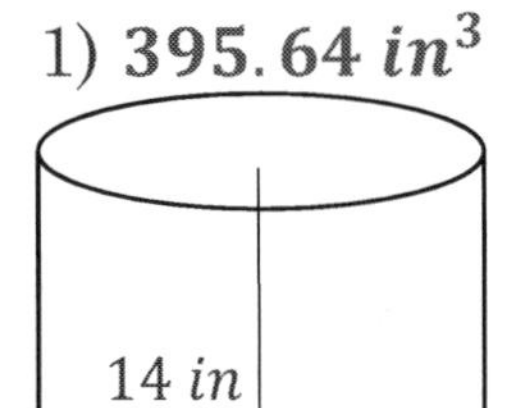

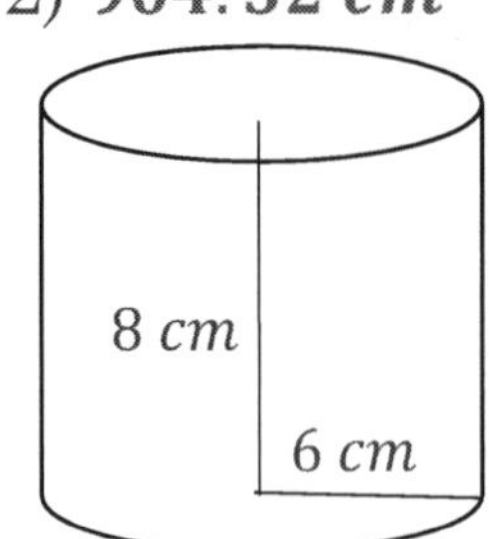

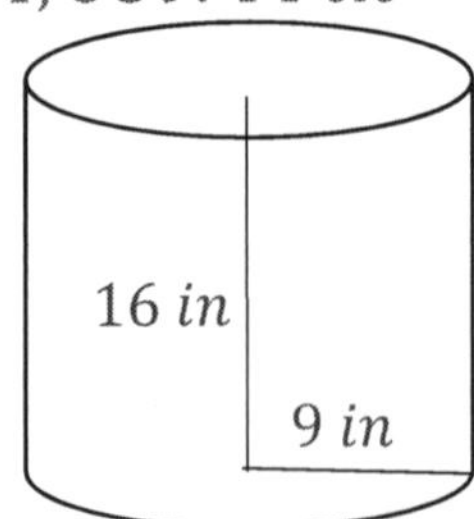

4) $\mathbf{4,019.2}\ ft^3$

5) 3,617.28 in^3

6) 9,947.52 in^3

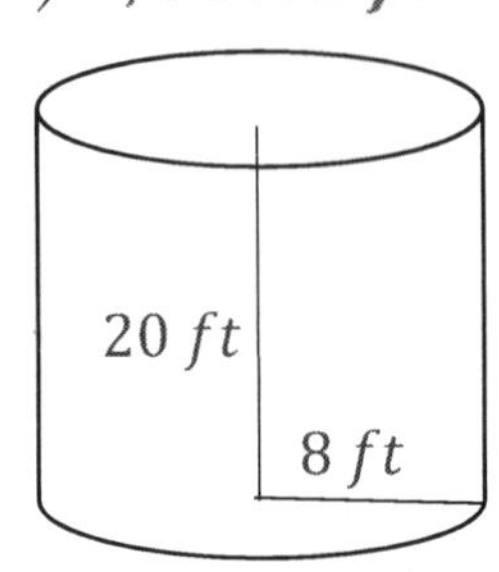

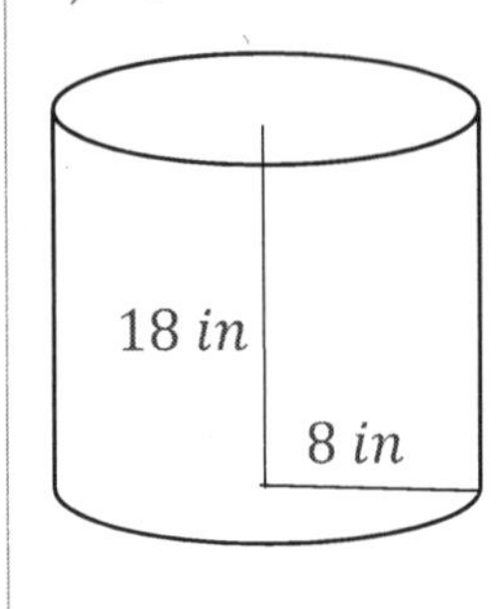

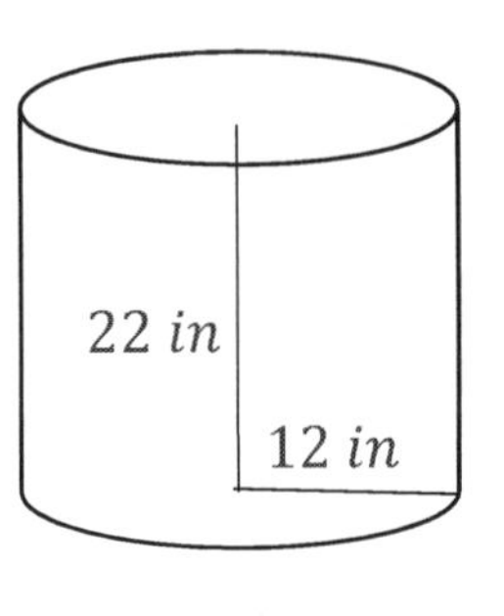

Find the surface area of each Cylinder. (π = 3.14)

7) 471 in^2

8) 301.44 cm^2

9) 533.8 ft^2

10) 502.4 m^2

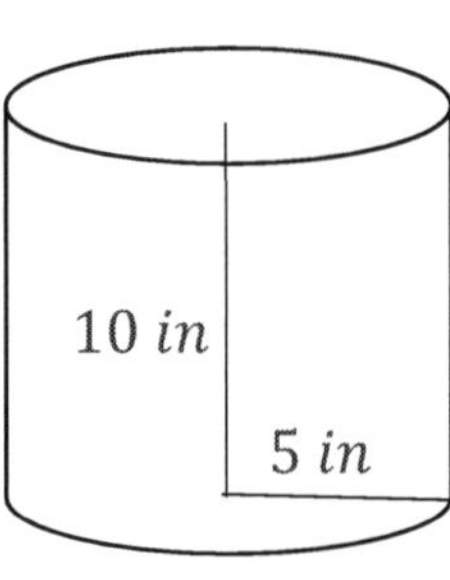

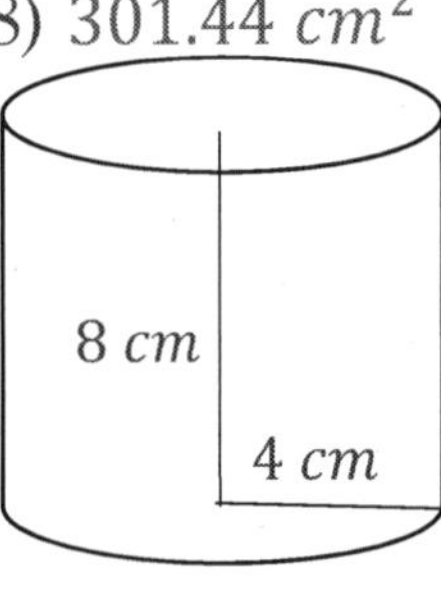

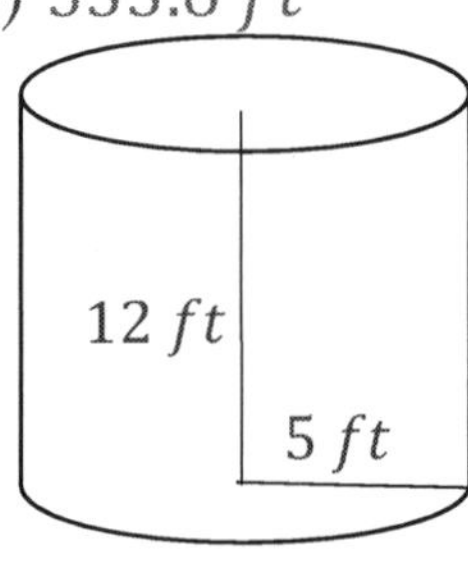

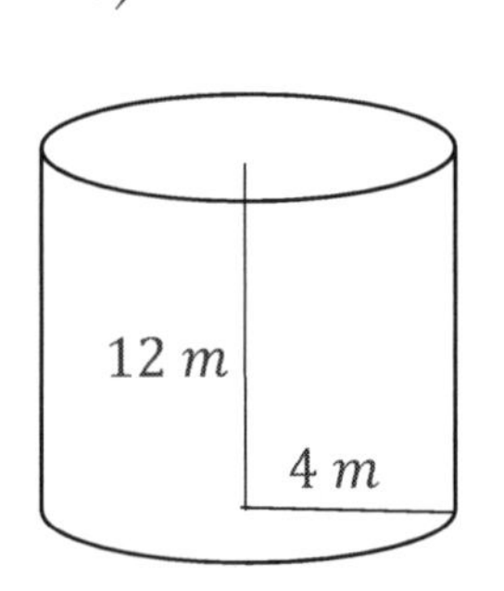

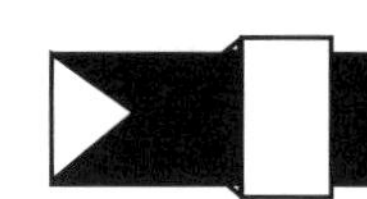

Mean, Median, Mode, and Range of the Given Data

Find the values of the Given Data.

1) 6, 12, 1, 1, 5

Mode: _____ Range: _____

Mean: _____ Median: _____

2) 5, 8, 3, 7, 4, 3

Mode: _____ Range: _____

Mean: _____ Median: _____

3) 12, 5, 8, 7, 8

Mode: _____ Range: _____

Mean: _____ Median: _____

4) 8, 4, 10, 7, 3, 4

Mode: _____ Range: _____

Mean: _____ Median: _____

5) 9, 7, 10, 5, 7, 4, 14

Mode: _____ Range: _____

Mean: _____ Median: _____

6) 8, 1, 6, 6, 9, 2, 17

Mode: _____ Range: _____

Mean: _____ Median: _____

7) 12, 6, 1, 7, 9, 7, 8, 14

Mode: _____ Range: _____

Mean: _____ Median: _____

8) 10, 14, 5, 4, 11, 6, 13

Mode: _____ Range: _____

Mean: _____ Median: _____

9) 16, 15, 15, 16, 13, 14, 23

Mode: _____ Range: _____

Mean: _____ Median: _____

10) 16, 15, 12, 8, 4, 9, 8, 16

Mode: _____ Range: _____

Mean: ____ Median: _____

Mean, Median, Mode, and Range of the Given Data – Answers

Find the values of the Given Data.

1) $6, 12, 1, 1, 5$

Mode: 1 Range: 11

Mean: 5 Median: 5

2) $5, 8, 3, 7, 4, 3$

Mode: 3 Range: 5

Mean: 5 Median: 4.5

3) $12, 5, 8, 7, 8$

Mode: 8 Range: 7

Mean: 8 Median: 8

4) $8, 4, 10, 7, 3, 4$

Mode: 4 Range: 7

Mean: 6 Median: 5.5

5) $9, 7, 10, 5, 7, 4, 14$

Mode: 7 Range: 10

Mean: 8 Median: 7

6) $8, 1, 6, 6, 9, 2, 17$

Mode: 6 Range: 16

Mean: 7 Median: 6

7) $12, 6, 1, 7, 9, 7, 8, 14$

Mode: 7 Range: 13

Mean: 8 Median: 7.5

8) $10, 14, 5, 4, 11, 6, 13$

Mode: *no mode* Range: 10

Mean: 9 Median: 10

9) $16, 15, 15, 16, 13, 14, 23$

Mode: 15 *and* 16 Range: 10

Mean: 16 Median: 15

10) $16, 15, 12, 8, 4, 9, 8, 16$

Mode: 8 *and* 16 Range: 12

Mean: 11 Median: 10.5

Pie Graph

The circle graph below shows all Wilson's expenses for last month. Wilson spent $200 on his bills last month.

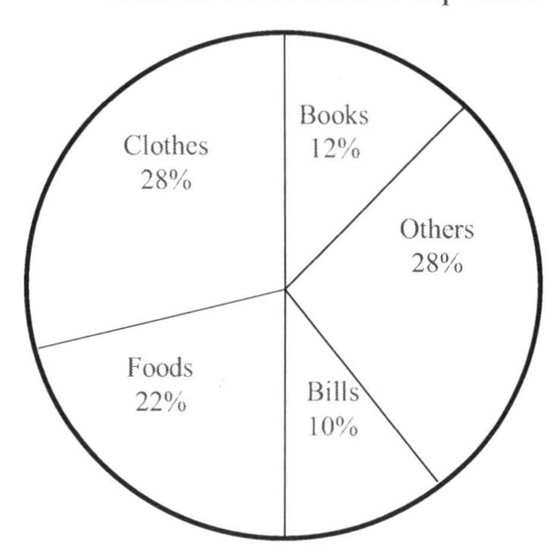

Answer following questions based on the Pie graph.

1) How much was Wilson's total expenses last month? ____________

2) How much did Wilson spend on his clothes last month? __________

3) How much did Wilson spend on foods last month? ______________

4) How much did Wilson spend on his books last month? _________

5) What fraction is Wilson's expenses for his bills and clothes out of his total expenses last month? ___________________

Pie Graph - Answers

The circle graph below shows all Wilson's expenses for last month. Wilson spent $200 on his bills last month.

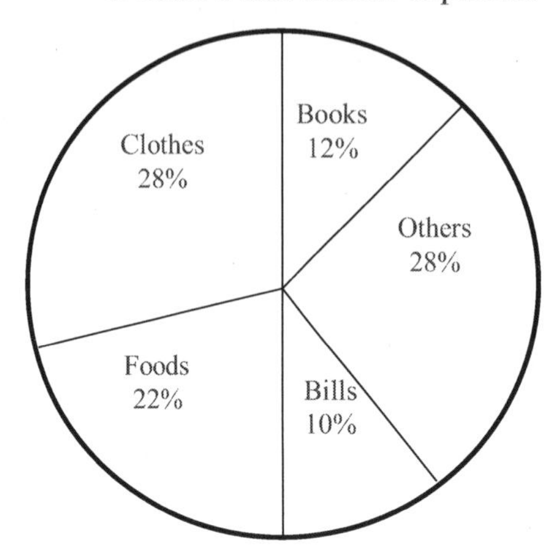

Answer following questions based on the Pie graph.

1) How much was Wilson's total expenses last month? $2,000

2) How much did Wilson spend on his clothes last month? $560

3) How much did Wilson spend on foods last month? $440

4) How much did Wilson spend on his books last month? $240

5) What fraction is Wilson's expenses for his bills and clothes out of his total expenses last month? $\frac{19}{50}$

Probability Problems

1) If there are 10 red balls and 20 blue balls in a basket, what is the probability that Oliver will pick out a red ball from the basket? ____________

Gender	Under 45	45 or older	total
Male	12	6	18
Female	5	7	12
Total	17	13	30

2) The table above shows the distribution of age and gender for 30 employees in a company. If one employee is selected at random, what is the probability that the employee selected be either a female under age 45 or a male age 45 or older? ____________

3) A number is chosen at random from 1 to 18. Find the probability of not selecting a composite number. (A composite number is a number that is divisible by itself, 1 and at least one other whole number) ____________

4) There are 6 blue marbles, 8 red marbles, and 5 yellow marbles in a box. If Ava randomly selects a marble from the box, what is the probability of selecting a red or yellow marble? ____________

5) A bag contains 19 balls: three green, five black, eight blue, a brown, a red and one white. If 18 balls are removed from the bag at random, what is the probability that a brown ball has been removed? ____________

6) There are only red and blue marbles in the box. The probability of randomly choosing a red marble in the box is one-fourth. If there are 132 blue marbles, how many are in the box?

____

Probability Problems - Answers

1) If there are 10 red balls and 20 blue balls in a basket, what is the probability that Oliver will pick out a red ball from the basket? $\frac{1}{3}$

Gender	Under 45	45 or older	total
Male	12	6	18
Female	5	7	12
Total	17	13	30

2) The table above shows the distribution of age and gender for 30 employees in a company. If one employee is selected at random, what is the probability that the employee selected be either a female under age 45 or a male age 45 or older? $\frac{11}{30}$

3) A number is chosen at random from 1 to 18. Find the probability of not selecting a composite number. (A composite number is a number that is divisible by itself, 1 and at least one other whole number) $\frac{7}{18}$

4) There are 6 blue marbles, 8 red marbles, and 5 yellow marbles in a box. If Ava randomly selects a marble from the box, what is the probability of selecting a red or yellow marble? $\frac{13}{19}$

5) A bag contains 19 balls: three green, five black, eight blue, a brown, a red and one white. If 18 balls are removed from the bag at random, what is the probability that a brown ball has been removed? $\frac{18}{19}$

6) There are only red and blue marbles in the box. The probability of randomly choosing a red marble in the box is one-fourth. If there are 132 blue marbles, how many are in the box?176

Permutations and Combinations

Calculate the value of each.

1) $5! =$ ____
2) $6! =$ ____
3) $8! =$ ____
4) $5! + 6! =$ ____
5) $8! + 3! =$ ____
6) $6! + 7! =$ ____
7) $8! + 4! =$ ____
8) $9! - 3! =$ ____

Solve each word problems.

9) Sophia is baking cookies. She uses milk, flour and eggs. How many different orders of ingredients can she try? _____

10) William is planning for his vacation. He wants to go to the restaurant, watch a movie, go to the beach, and play basketball. How many different ways of ordering are there for him? _____

11) How many 7-digit numbers can be named using the digits $1, 2, 3, 4, 5, 6$ and 7 without repetition? _____

12) In how many ways can 9 boys be arranged in a straight line? _____

13) In how many ways can 10 athletes be arranged in a straight line? _____

14) A professor is going to arrange her 7 students in a straight line. In how many ways can she do this? _____

15) How many code symbols can be formed with the letters for the word BLACK? _____

16) In how many ways a team of 7 basketball players can choose a captain and co-captain? _____

Permutations and Combinations - Answers

Calculate the value of each.

1) $5! = 120$
2) $6! = 720$
3) $8! = 40{,}320$
4) $5! + 6! = 840$
5) $8! + 3! = 40{,}326$
6) $6! + 7! = 5{,}760$
7) $8! + 4! = 40{,}344$
8) $9! - 3! = 362{,}874$

Solve each word problems.

9) Sophia is baking cookies. She uses milk, flour and eggs. How many different orders of ingredients can she try? 6
10) William is planning for his vacation. He wants to go to the restaurant, watch a movie, go to the beach, and play basketball. How many different ways of ordering are there for him? 24
11) How many 7-digit numbers can be named using the digits $1, 2, 3, 4, 5, 6$ and 7 without repetition? 5,040
12) In how many ways can 9 boys be arranged in a straight line? 362,880
13) In how many ways can 10 athletes be arranged in a straight line? 3,628,800
14) A professor is going to arrange her 7 students in a straight line. In how many ways can she do this? 5,040
15) How many code symbols can be formed with the letters for the word BLACK? 120
16) In how many ways a team of 7 basketball players can choose a captain and co-captain? 42

Function Notation and Evaluation

Evaluate each function.

1) $f(x) = x - 1$, find $f(-1)$

_____

2) $g(x) = x + 3$, find $g(4)$

_____

3) $h(x) = x + 9$, find $h(3)$

_____

4) $f(x) = -x - 6$, find $f(5)$

_____

5) $f(x) = 2x - 7$, find $f(-1)$

_____

6) $w(x) = -2 - 4x$, find $w(5)$

_____

7) $g(n) = 6n - 3$, find $g(-2)$

_____

8) $h(x) = -8x + 12$, find $h(3)$

_____

9) $k(n) = 14 - 3n$, find $k(3)$

_____

10) $g(x) = 4x - 4$, find $g(-2)$

_____

11) $k(n) = 8n - 7$, find $k(4)$

_____

12) $w(n) = -2n + 14$, find $w(5)$

_____

13) $h(x) = 5x - 18$, find $h(8)$

_____

14) $g(n) = 2n^2 + 2$, find $g(5)$

_____

15) $f(x) = 3x^2 - 13$, find $f(2)$

_____

16) $g(n) = 5n^2 + 7$, find $g(-3)$

_____

17) $h(n) = 5n^2 - 10$, find $h(4)$

_____

18) $g(x) = -3x^2 - 6x$, find $g(2)$

_____

19) $k(n) = 3n^3 + 2n$, find $k(-5)$

_____

20) $f(x) = -4x + 12$, find $f(2x)$

_____

21) $k(a) = 6a + 5$, find $k(a - 1)$

_____

22) $h(x) = 9x + 3$, find $h(5x)$

_____

Function Notation and Evaluation – Answer

Evaluate each function.

1) $f(x) = x - 1$, find $f(-1)$

   -2

2) $g(x) = x + 3$, find $g(4)$

   7

3) $h(x) = x + 9$, find $h(3)$

   12

4) $f(x) = -x - 6$, find $f(5)$

   -11

5) $f(x) = 2x - 7$, find $f(-1)$

   -9

6) $w(x) = -2 - 4x$, find $w(5)$

   -22

7) $g(n) = 6n - 3$, find $g(-2)$

   -15

8) $h(x) = -8x + 12$, find $h(3)$

   -12

9) $k(n) = 14 - 3n$, find $k(3)$

   5

10) $g(x) = 4x - 4$, find $g(-2)$

   -12

11) $k(n) = 8n - 7$, find $k(4)$

   25

12) $w(n) = -2n + 14$, find $w(5)$

   4

13) $h(x) = 5x - 18$, find $h(8)$

   22

14) $g(n) = 2n^2 + 2$, find $g(5)$

   52

15) $f(x) = 3x^2 - 13$, find $f(2)$

   -1

16) $g(n) = 5n^2 + 7$, find $g(-3)$

   52

17) $h(n) = 5n^2 - 10$, find $h(4)$

   70

18) $g(x) = -3x^2 - 6x$, find $g(2)$

   -24

19) $k(n) = 3n^3 + 2n$, find $k(-5)$

   -385

20) $f(x) = -4x + 12$, find $f(2x)$

   $-8x + 12$

21) $k(a) = 6a + 5$, find $k(a - 1)$

   $6a - 1$

22) $h(x) = 9x + 3$, find $h(5x)$

   $45x + 3$

Adding and Subtracting Functions

Perform the indicated operation.

1) $f(x) = x + 6$

   $g(x) = 3x + 3$

   Find $(f - g)(2)$

   ______

2) $g(x) = x - 3$

   $f(x) = -x - 4$

   Find $(g - f)(-2)$

   ______

3) $h(t) = 5t + 4$

   $g(t) = 2t + 2$

   Find $(h + g)(-1)$

   ______

4) $g(a) = 3a - 5$

   $f(a) = a^2 + 6$

   Find $(g + f)(3)$

   ______

5) $g(x) = 4x - 5$

   $h(x) = 6x^2 + 5$

   Find $(g - f)(-2)$

   ______

6) $h(x) = x^2 + 3$

   $g(x) = -4x + 1$

   Find $(h + g)(4)$

   ______

7) $f(x) = -2x - 8$

   $g(x) = x^2 + 2$

   Find $(f - g)(6)$

   ______

8) $h(n) = -4n^2 + 9$

   $g(n) = 5n + 6$

   Find $(h - g)(5)$

   ______

9) $g(x) = 3x^2 - 2x - 1$

   $f(x) = 5x + 12$

   Find $(g - f)(a)$

   ______

10) $g(t) = -5t - 8$

   $f(t) = -t^2 + 2t + 12$

   Find $(g + f)(x)$

   ______

Adding and Subtracting Functions - answer

Perform the indicated operation.

1) $f(x) = x + 6$

$g(x) = 3x + 3$

Find $(f - g)(2)$

-1

2) $g(x) = x - 3$

$f(x) = -x - 4$

Find $(g - f)(-2)$

-3

3) $h(t) = 5t + 4$

$g(t) = 2t + 2$

Find $(h + g)(-1)$

-1

4) $g(a) = 3a - 5$

$f(a) = a^2 + 6$

Find $(g + f)(3)$

19

5) $g(x) = 4x - 5$

$h(x) = 6x^2 + 5$

Find $(g - f)(-2)$

-42

6) $h(x) = x^2 + 3$

$g(x) = -4x + 1$

Find $(h + g)(4)$

4

7) $f(x) = -2x - 8$

$g(x) = x^2 + 2$

Find $(f - g)(6)$

-58

8) $h(n) = -4n^2 + 9$

$g(n) = 5n + 6$

Find $(h - g)(5)$

-122

9) $g(x) = 3x^2 - 2x - 1$

$f(x) = 5x + 12$

Find $(g - f)(a)$

$3a^2 - 7a - 13$

10) $g(t) = -5t - 8$

$f(t) = -t^2 + 2t + 12$

Find $(g + f)(x)$

$-x^2 - 3x + 4$

Multiplying and Dividing Functions

Perform the indicated operation.

1) $g(x) = x + 2$

   $f(x) = x + 3$

   Find $(g.f)(4)$

   ______

2) $g(a) = a + 2$

   $h(a) = 2a - 3$

   Find $(g.h)(5)$

   ______

3) $f(x) = a^2 - 2$

   $g(x) = -4 + 3a$

   Find $(fg)(2)$

   ______

4) $g(t) = t^2 + 4$

   $h(t) = 2t - 4$

   Find $(g.h)(-3)$

   ______

5) $g(a) = 2a^2 - 4a + 2$

   $f(a) = 2a^3 - 2$

   Find $(\frac{g}{f})(4)$

   ______

6) $f(x) = 2x$

   $h(x) = -x + 6$

   Find $(f.h)(-2)$

   ______

7) $f(x) = 2x + 4$

   $h(x) = 4x - 2$

   Find $(\frac{f}{h})(2)$

   ______

8) $g(a) = 4a + 6$

   $f(a) = 2a - 8$

   Find $(\frac{g}{f})(3)$

   ______

9) $g(x) = x^2 + 2x + 5$

   $h(x) = 2x + 3$

   Find $(g.h)(2)$

   ______

10) $g(x) = -4x^2 + 5 - 2x$

    $f(x) = x^2 - 2$

    Find $(g.f)(3)$

    ______

Multiplying and Dividing Functions - answer

Perform the indicated operation.

1) $g(x) = x + 2$

$f(x) = x + 3$

Find $(g.f)(4)$

42

2) $g(a) = a + 2$

$h(a) = 2a - 3$

Find $(g.h)(5)$

49

3) $f(x) = a^2 - 2$

$g(x) = -4 + 3a$

Find $(fg)(2)$

1

4) $g(t) = t^2 + 4$

$h(t) = 2t - 4$

Find $(g.h)(-3)$

-130

5) $g(a) = 2a^2 - 4a + 2$

$f(a) = 2a^3 - 2$

Find $(\frac{g}{f})(4)$

$\frac{1}{7}$

6) $f(x) = 2x$

$h(x) = -x + 6$

Find $(f.h)(-2)$

-32

7) $f(x) = 2x + 4$

$h(x) = 4x - 2$

Find $(\frac{f}{h})(2)$

$\frac{4}{3}$

8) $g(a) = 4a + 6$

$f(a) = 2a - 8$

Find $(\frac{g}{f})(3)$

-9

9) $g(x) = x^2 + 2x + 5$

$h(x) = 2x + 3$

Find $(g.h)(2)$

91

10) $g(x) = -4x^2 + 5 - 2x$

$f(x) = x^2 - 2$

Find $(g.f)(3)$

-259

Composition of Functions

Using $f(x) = x + 4$ ***and*** $g(x) = 2x$***, find:***

1) $f(g(1)) =$ ____
2) $f(g(-1)) =$ ____
3) $g(f(-2)) =$ ____
4) $g(f(2)) =$ ____
5) $f(g(2)) =$ ____
6) $g(f(3)) =$ ____

Using $f(x) = 2x + 5$ ***and*** $g(x) = x - 2$***, find:***

7) $g(f(2)) =$ ____
8) $g(f(-2)) =$ ____
9) $f(g(5)) =$ ____
10) $f(f(4)) =$ ____
11) $g(f(3)) =$ ____
12) $g(f(-3)) =$ ____

Using $f(x) = 4x - 2$ ***and*** $g(x) = x - 5$***, find:***

13) $g(f(-2)) =$ ____
14) $f(f(4)) =$ ____
15) $f(g(5)) =$ ____
16) $f(f(3)) =$ ____
17) $g(f(-3)) =$ ____
18) $g(g(6)) =$ ____

Using $f(x) = 5x + 3$ ***and*** $g(x) = 2x - 5$***, find:***

19) $f(g(-4)) =$ ____
20) $g(f(6)) =$ ____
21) $f(g(5)) =$ ____
22) $f(f(3)) =$ ____

Composition of Functions - Answer

Using $f(x) = x + 4$ ***and*** $g(x) = 2x$***, find:***

1) $f(g(1)) = 6$
2) $f(g(-1)) = 2$
3) $g(f(-2)) = 4$
4) $g(f(2)) = 12$
5) $f(g(2)) = 8$
6) $g(f(3)) = 14$

Using $f(x) = 2x + 5$ ***and*** $g(x) = x - 2$***, find:***

7) $g(f(2)) = 7$
8) $g(f(-2)) = -1$
9) $f(g(5)) = 11$
10) $f(f(4)) = 31$
11) $g(f(3)) = 9$
12) $g(f(-3)) = -3$

Using $f(x) = 4x - 2$ ***and*** $g(x) = x - 5$***, find:***

13) $g(f(-2)) = -15$
14) $f(f(4)) = 54$
15) $f(g(5)) = -2$
16) $f(f(3)) = 38$
17) $g(f(-3)) = -19$
18) $g(g(6)) = -4$

Using $f(x) = 5x + 3$ ***and*** $g(x) = 2x - 5$***, find:***

19) $f(g(-4)) = -62$
20) $g(f(6)) = 61$
21) $f(g(5)) = 28$
22) $f(f(3)) = 93$

Solving a Quadratic Equation

Solve each equation by factoring or using the quadratic formula.

1) $x^2 - 4x - 32 = 0$

____

2) $x^2 - 2x - 63 = 0$

____

3) $x^2 + 17x + 72 = 0$

____

4) $x^2 + 14x + 48 = 0$

____

5) $x^2 + 5x - 24 = 0$

____

6) $x^2 + 15x + 36 = 0$

____

7) $x^2 + 12x - 28 = 0$

____

8) $x^2 + 6x - 55 = 0$

____

9) $x^2 + 16x - 105 = 0$

____

10) $x^2 - 21x + 54 = 0$

____

11) $x^2 + 8x - 128 = 0$

____

12) $x^2 + 19x - 150 = 0$

____

13) $x^2 + 15x - 154 = 0$

____

14) $2x^2 - 2x - 60 = 0$

____

15) $2x^2 - 10x - 72 = 0$

____

16) $4x^2 + 48x + 128 = 0$

____

17) $4x^2 + 40x + 96 = 0$

____

18) $2x^2 + 28x + 90 = 0$

____

19) $9x^2 + 63x + 108 = 0$

____

20) $4x^2 + 56x + 160 = 0$

____

Solving a Quadratic Equation - Answers

Solve each equation by factoring or using the quadratic formula.

1) $x^2 - 4x - 32 = 0$

$x = 8, x = -4$

2) $x^2 - 2x - 63 = 0$

$x = 9, x = -7$

3) $x^2 + 17x + 72 = 0$

$x = -9, x = -8$

4) $x^2 + 14x + 48 = 0$

$x = -6, x = -8$

5) $x^2 + 5x - 24 = 0$

$x = 3, x = -8$

6) $x^2 + 15x + 36 = 0$

$x = -12, x = -3$

7) $x^2 + 12x - 28 = 0$

$x = 2, x = -14$

8) $x^2 + 6x - 55 = 0$

$x = 5, x = -11$

9) $x^2 + 16x - 105 = 0$

$x = -21, x = 5$

10) $x^2 - 21x + 54 = 0$

$x = 18, x = 3$

11) $x^2 + 8x - 128 = 0$

$x = -16, x = 8$

12) $x^2 + 19x - 150 = 0$

$x = -25, x = 6$

13) $x^2 + 15x - 154 = 0$

$x = -22, x = 7$

14) $2x^2 - 2x - 60 = 0$

$x = 6, x = -5$

15) $2x^2 - 10x - 72 = 0$

$x = 9, x = -4$

16) $4x^2 + 48x + 128 = 0$

$x = -4, x = -8$

17) $4x^2 + 40x + 96 = 0$

$x = -4, x = -6$

18) $2x^2 + 28x + 90 = 0$

$x = -5, x = -9$

19) $9x^2 + 63x + 108 = 0$

$x = -3, x = -4$

20) $4x^2 + 56x + 160 = 0$

$x = -4, x = -10$

Graphing Quadratic Functions

Sketch the graph of each function.

1) $y = (x+1)^2 - 2$

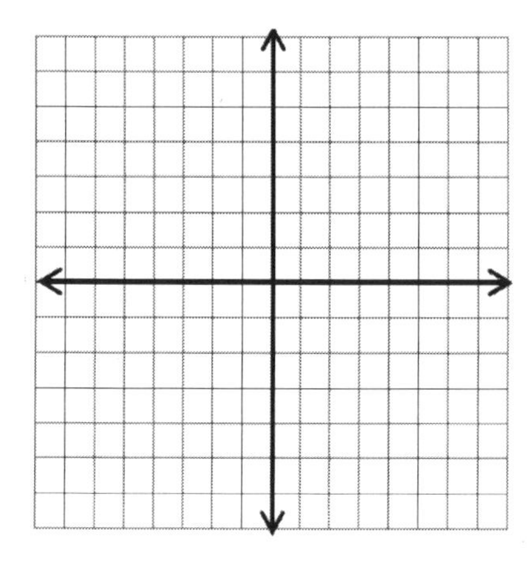

2) $y = (x-1)^2 + 3$

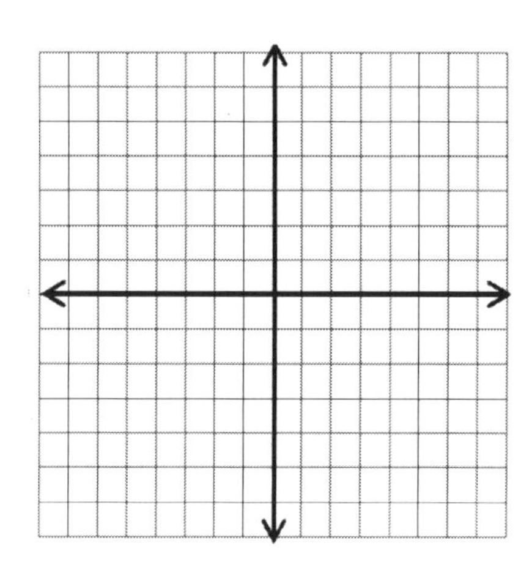

3) $y = x^2 - 4x + 6$

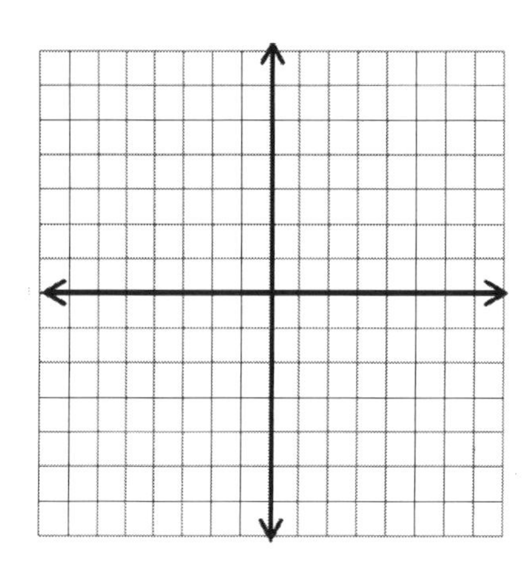

4) $y = x^2 - 6x + 14$

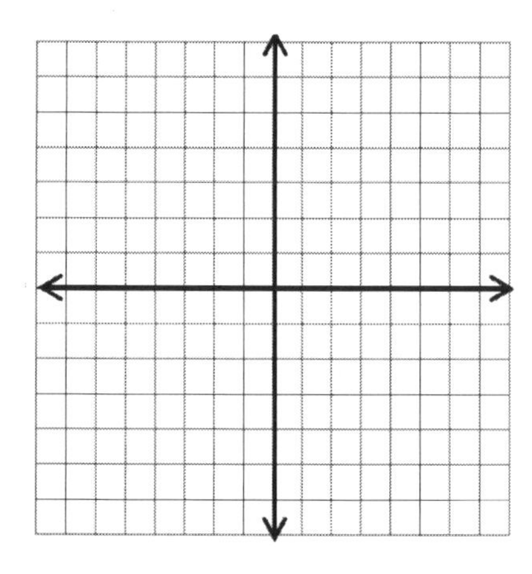

5) $y = x^2 + 12x + 34$

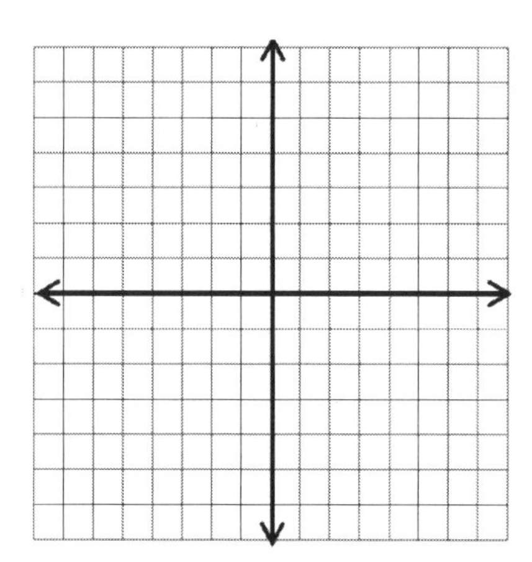

6) $y = 2(x+1)^2 - 4$

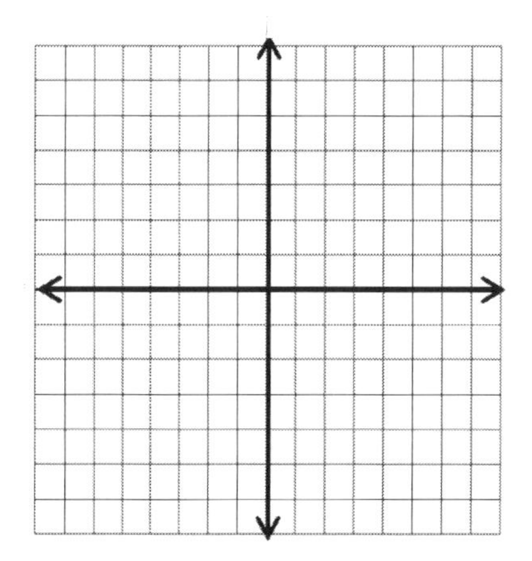

Graphing Quadratic Functions - Answers

Sketch the graph of each function.

1) $y = (x + 1)^2 - 2$

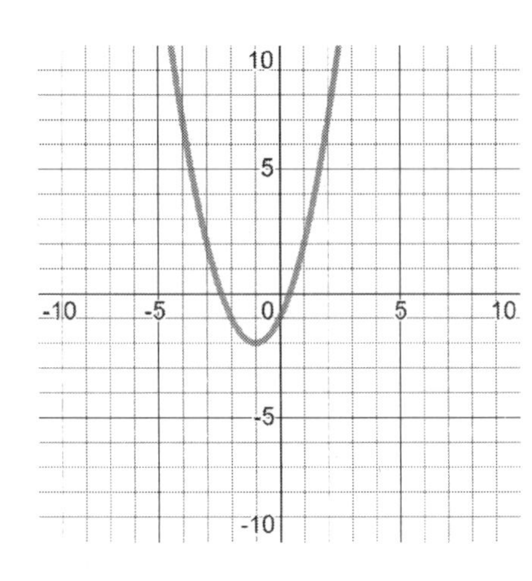

2) $y = (x - 1)^2 + 3$

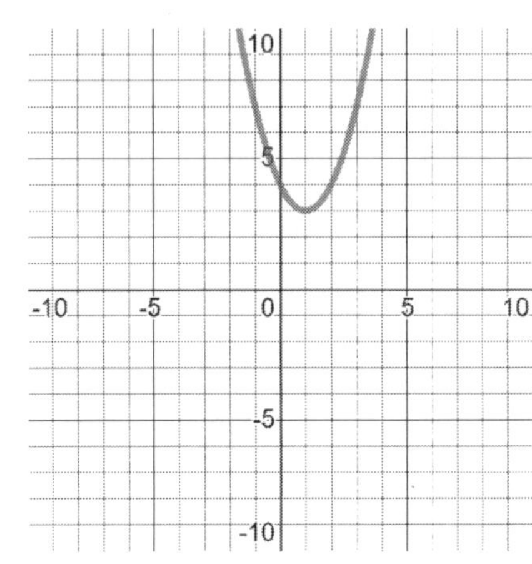

3) $y = x^2 - 4x + 6$

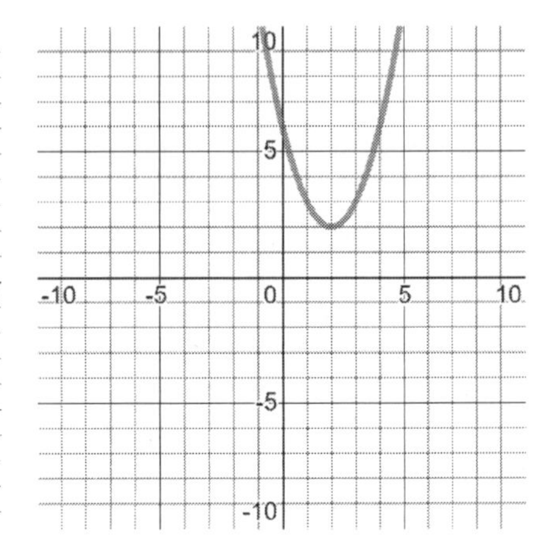

4) $y = x^2 - 6x + 14$

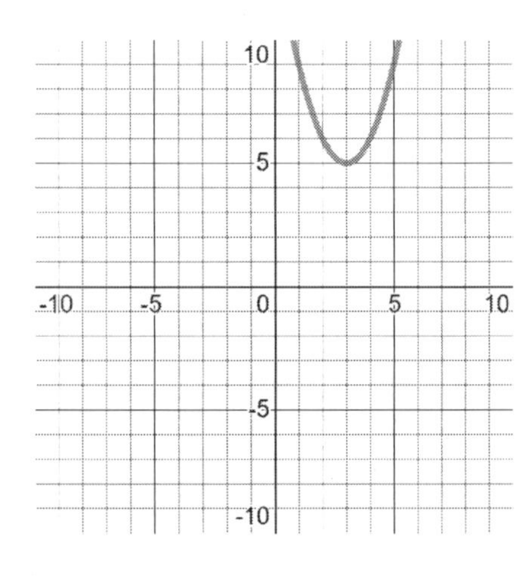

5) $y = x^2 + 12x + 34$

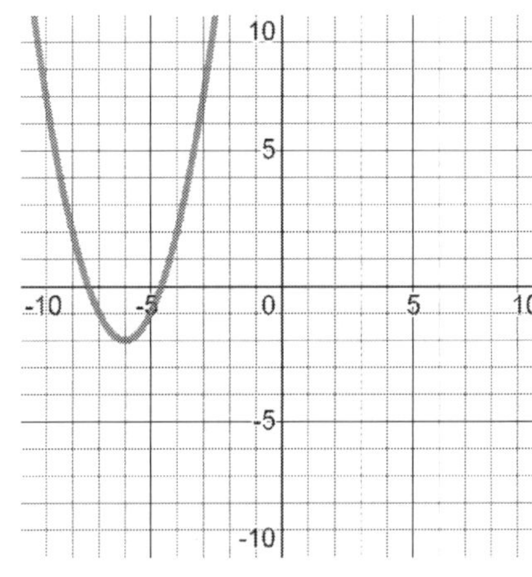

6) $y = 2(x + 1)^2 - 4$

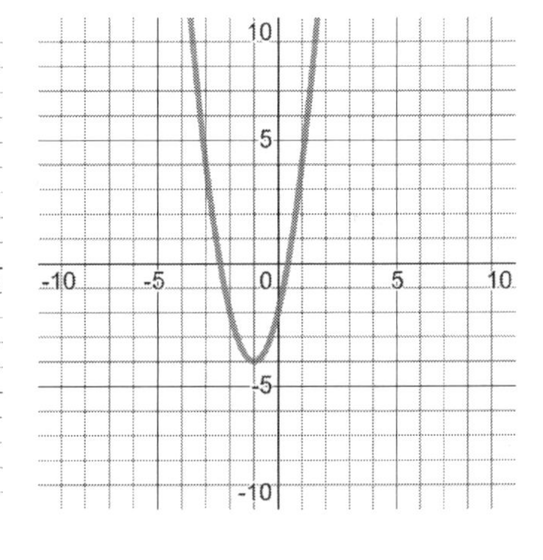

Solving Quadratic Inequalities

Solve each quadratic inequality.

1) $x^2 - 4 < 0$

_____

2) $x^2 - 9 > 0$

_____

3) $x^2 - 5x - 6 < 0$

_____

4) $x^2 + 8x - 20 > 0$

_____

5) $x^2 + 10x - 24 \geq 0$

_____

6) $x^2 - 15x + 54 < 0$

_____

7) $x^2 + 17x + 72 \leq 0$

_____

8) $x^2 + 15x + 44 \geq 0$

_____

9) $x^2 + 5x - 50 \geq 0$

_____

10) $x^2 - 18x + 72 < 0$

_____

11) $x^2 - 18x + 45 > 0$

_____

12) $x^2 + 16x - 80 > 0$

_____

13) $x^2 + 9x - 112 \leq 0$

_____

14) $x^2 + 4x - 117 \leq 0$

_____

15) $x^2 + 19x + 88 \geq 0$

_____

16) $x^2 + 26x + 168 \leq 0$

_____

17) $4x^2 + 24x + 32 \leq 0$

_____

18) $4x^2 - 4x - 48 \geq 0$

_____

19) $4x^2 - 16x - 48 \leq 0$

_____

20) $9x^2 - 63x + 108 > 0$

_____

Solving Quadratic Inequalities- Answers

Solve each quadratic inequality.

1) $x^2 - 4 < 0$

$-2 < x < 2$

2) $x^2 - 9 > 0$

$-3 < x < 3$

3) $x^2 - 5x - 6 < 0$

$-1 < x < 6$

4) $x^2 + 8x - 20 > 0$

$x < -10 \text{ or } x > 2$

5) $x^2 + 10x - 24 \geq 0$

$x \leq -12 \text{ or } x \geq 2$

6) $x^2 - 15x + 54 < 0$

$6 < x < 9$

7) $x^2 + 17x + 72 \leq 0$

$-9 \leq x \leq -8$

8) $x^2 + 15x + 44 \geq 0$

$x \leq -11 \text{ or } x \geq -4$

9) $x^2 + 5x - 50 \geq 0$

$x \leq -10 \text{ or } x \geq 5$

10) $x^2 - 18x + 72 < 0$

$6 \leq x \leq 12$

11) $x^2 - 18x + 45 > 0$

$x < 3 \text{ or } x > 15$

12) $x^2 + 16x - 80 > 0$

$x < -20 \text{ or } x > 4$

13) $x^2 + 9x - 112 \leq 0$

$-16 \leq x \leq 7$

14) $x^2 + 4x - 117 \leq 0$

$-13 \leq x \leq 9$

15) $x^2 + 19x + 88 \geq 0$

$x \leq -11 \text{ or } x \geq -8$

16) $x^2 + 26x + 168 \leq 0$

$-14 \leq x \leq -12$

17) $4x^2 + 24x + 32 \leq 0$

$-4 \leq x \leq -2$

18) $4x^2 - 4x - 48 \geq 0$

$x \leq -3 \text{ or } x \geq 4$

19) $4x^2 - 16x - 48 \leq 0$

$-2 \leq x \leq 6$

20) $9x^2 - 63x + 108 > 0$

$x \leq 3 \text{ or } x \geq 4$

Graphing Quadratic Inequalities

Sketch the graph of each quadratic inequality.

1) $y < -2x^2$

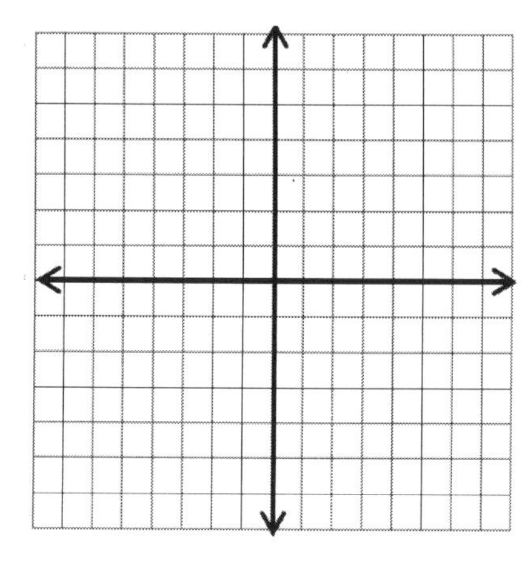

2) $y > 3x^2$

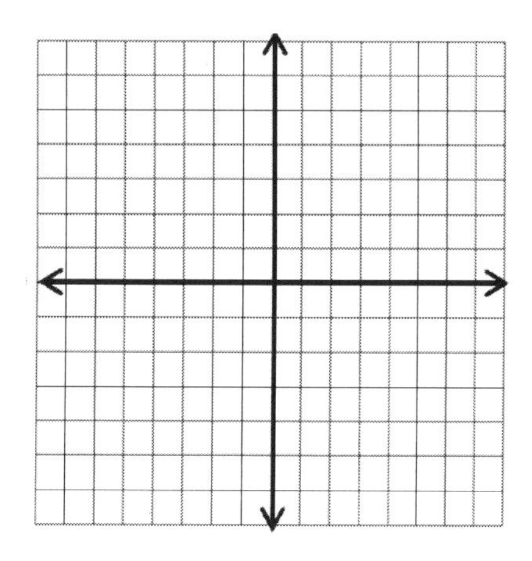

3) $y \geq -3x^2$

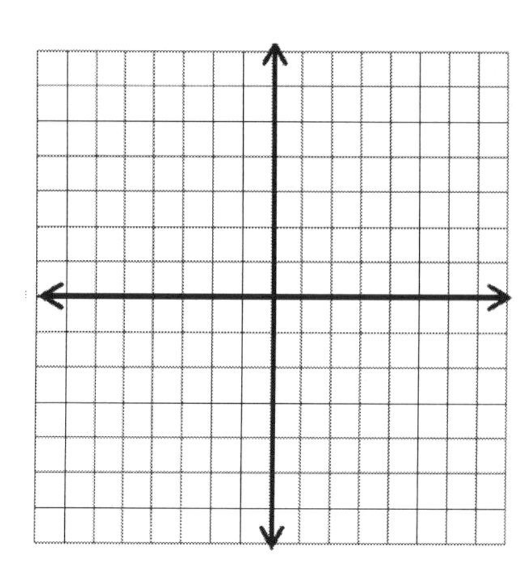

4) $y < x^2 + 1$

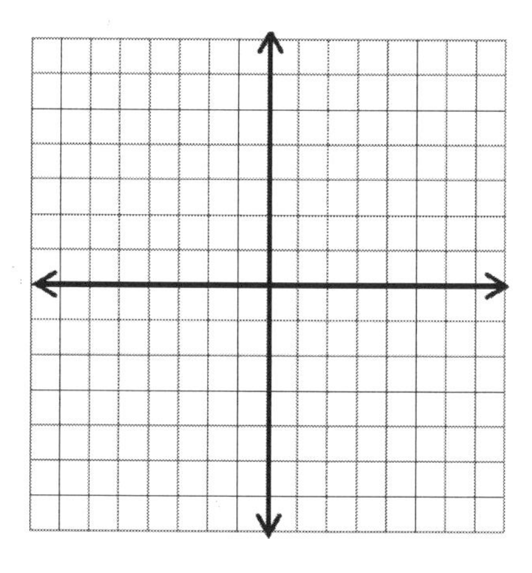

5) $y \geq -x^2 + 2$

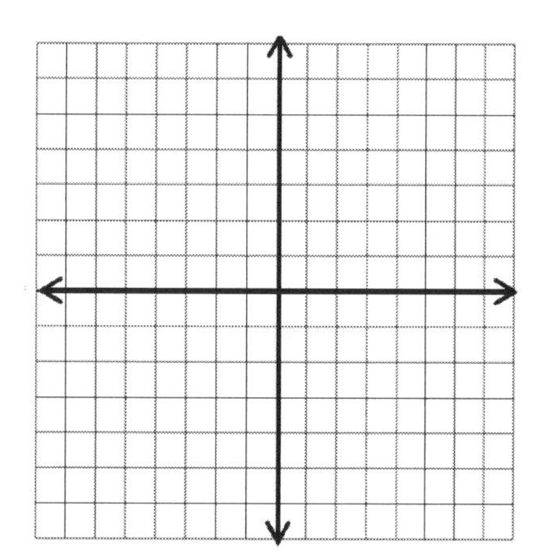

6) $y \leq x^2 - 2x - 3$

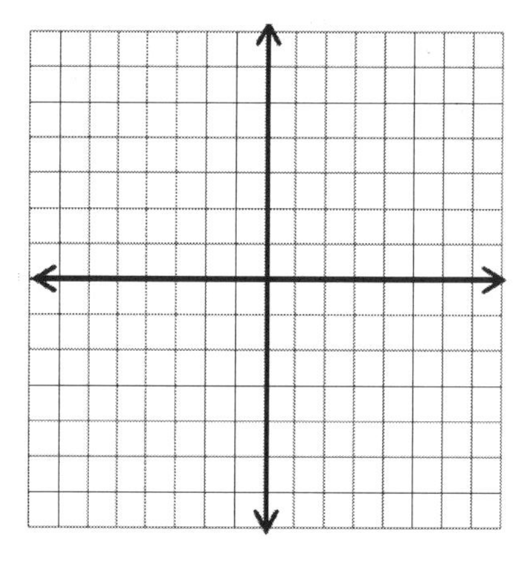

Graphing Quadratic Inequalities- Answers

Sketch the graph of each quadratic inequality.

1) $y < -2x^2$

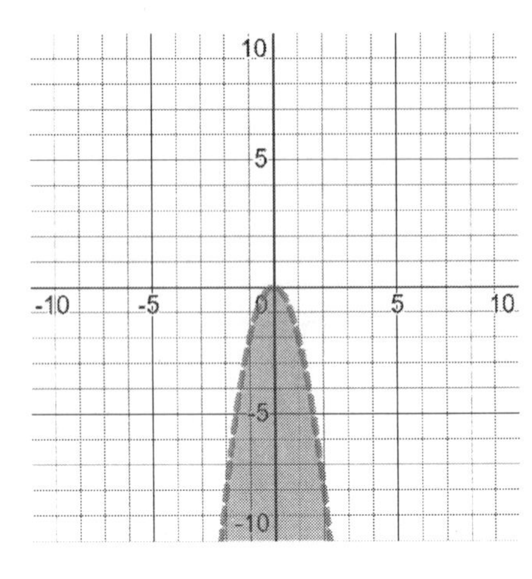

2) $y > 3x^2$

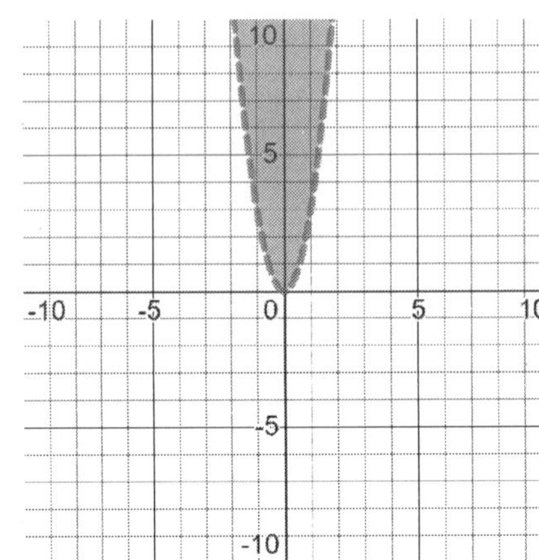

3) $y \geq -3x^2$

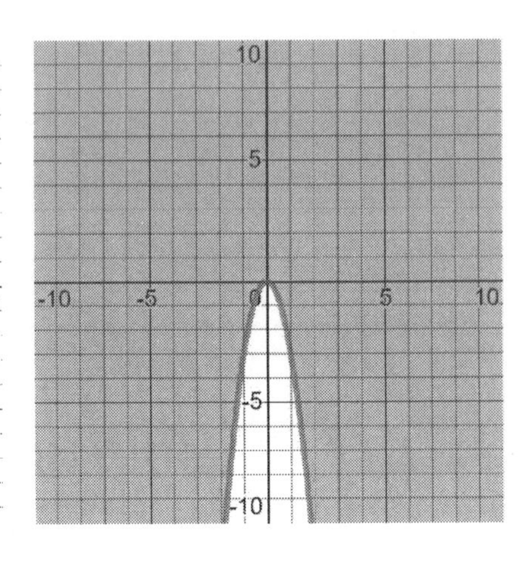

4) $y < x^2 + 1$

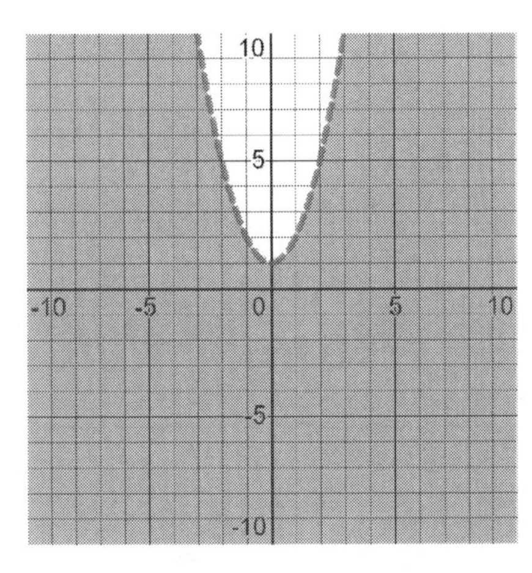

5) $y \geq -x^2 + 2$

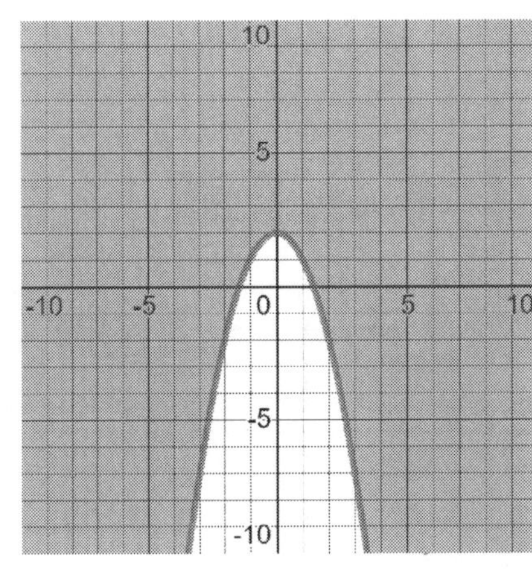

6) $y \leq x^2 - 2x - 3$

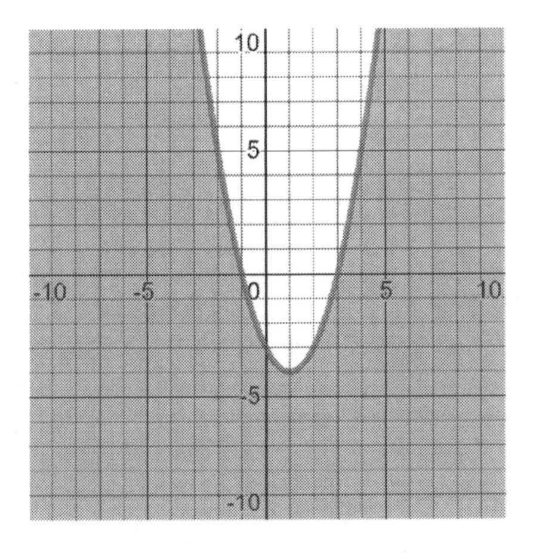

Cracking GACE Math Test

Embark on your journey to conquering the GACE Math exam!

Congratulations on reaching a vital milestone in your GACE Math preparation. With a solid understanding of the concepts under your belt, you're now ready to elevate your study by applying what you've learned. "***Cracking GACE Math Test***" is designed to bridge the gap between theory and practice, offering you a realistic testing experience.

Dive into the specifics of what this section offers:

GACE Math Test-Taking Strategies: Learn techniques that can help streamline your problem-solving process and enhance your accuracy under timed conditions.

GACE Math– Test Day Tips: Equip yourself with actionable advice to maintain calm and focus when it counts.

Now, put your knowledge into action:

GACE Math Practice Test 1 and 2: Step into the test-taker's shoes with practice tests crafted to mirror the actual exam's format and rigor.

GACE Math Practice Tests Answer Keys: Verify your solutions and understand your strengths and areas for improvement.

Answers and Explanations for Practice Tests 1 and 2: Benefit from detailed explanations to deepen your comprehension and correct misunderstandings.

Each practice question you work through is an opportunity to refine your skills and build confidence. Embrace this phase as a key part of your journey towards GACE Math success.

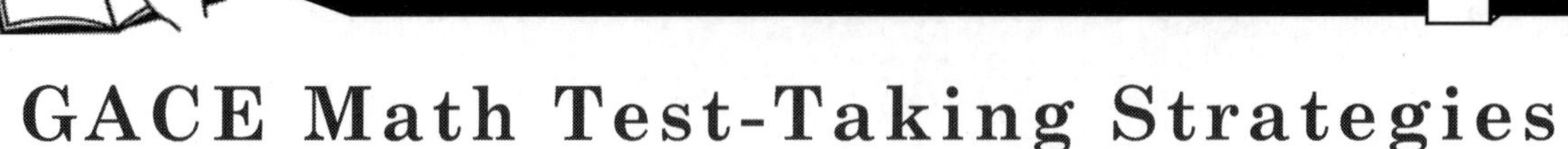

GACE Math Test-Taking Strategies

Here are some test-taking strategies that you can use to maximize your performance and results on the GACE Math test.

#1: Use This Approach To Answer Every GACE Math Question

- Review the question to identify keywords and important information.
- Translate the keywords into math operations so you can solve the problem.
- Review the answer choices. What are the differences between answer choices?
- Draw or label a diagram if needed.
- Try to find patterns.
- Find the right method to answer the question. Use straightforward math, plug in numbers, or test the answer choices (backsolving).
- Double-check your work.

#2: Use Educated Guessing

This approach is applicable to the problems you understand to some degree but cannot solve using straightforward math. In such cases, try to filter out as many answer choices as possible before picking an answer. In cases where you don't have a clue about what a certain problem entails, don't waste any time trying to eliminate answer choices. Just choose one randomly before moving onto the next question.

As you can ascertain, direct solutions are the most optimal approach. Carefully read through the question, determine what the solution is using the math you have learned before, then coordinate the answer with one of the choices available to you. Are you stumped? Make your best guess, then move on.

Don't leave any fields empty! Even if you're unable to work out a problem, strive to answer it. Take a guess if you have to. You will not lose points by getting an answer wrong, though you may gain a point by getting it correct!

#3 : BALLPARK

A ballpark answer is a rough approximation. When we become overwhelmed by calculations and figures, we end up making silly mistakes. A decimal that is moved by one unit can change an answer from right to wrong, regardless of the number of steps that you went through to get it. That's where ballparking can play a big part.

If you think you know what the correct answer may be (even if it's just a ballpark answer), you'll usually have the ability to eliminate a couple of choices. While answer choices are usually based on the average student error and/or values that are closely tied, you will still be able to weed out choices that are way far afield. Try to find answers that aren't in the proverbial ballpark when you're looking for a wrong answer on a multiple-choice question. This is an optimal approach to eliminating answers to a problem.

#4 : BACKSOLVING

All questions on the GACE Math test will be in multiple-choice format. Many test-takers prefer multiple-choice questions, as at least the answer is right there. You'll typically have four answers to pick from. You simply need to figure out which one is correct. Usually, the best way to go about doing so is "backsolving."

As mentioned earlier, direct solutions are the most optimal approach to answering a question. Carefully read through a problem, calculate a solution, then correspond the answer with one of the choices displayed in front of you. If you can't calculate a solution, your next best approach involves "backsolving."

When backsolving a problem, contrast one of your answer options against the problem you are asked, then see which of them is most relevant. More often than not, answer choices are listed in ascending or descending order. In such cases, try out the choices B or C. If it's not correct, you can go either down or up from there.

#5 : Plugging In Numbers

"Plugging in numbers" is a strategy that can be applied to a wide range of different math problems on the GACE Math test. This approach is typically used to simplify a challenging question so that it is more understandable. By using the strategy carefully, you can find the answer without too much trouble.

The concept is fairly straightforward–replace unknown variables in a problem with certain values. When selecting a number, consider the following:

- Choose a number that's basic (just not too basic). Generally, you should avoid choosing 1 (or even 0). A decent choice is 2.
- Try not to choose a number that is displayed in the problem.
- Make sure you keep your numbers different if you need to choose at least two of them.
- More often than not, choosing numbers merely lets you filter out some of your answer choices. As such, don't just go with the first choice that gives you the right answer.
- If several answers seem correct, then you'll need to choose another value and try again. This time, though, you'll just need to check choices that haven't been eliminated yet.
- If your question contains fractions, then a potential right answer may involve either an LCD (least common denominator) or an LCD multiple.
- 100 is the number you should choose when you are dealing with problems involving percentages.

GACE Math– Daytime Tips

After practicing and reviewing all the math concepts you've been taught, and taking some GACE Math practice tests, you'll be prepared for test day. Consider the following tips to be extra-ready come test time.

Before Your Test

What to do the night before:

- **Relax!** One day before your test, study lightly or skip studying altogether. You shouldn't attempt to learn something new, either. There are plenty of reasons why studying the evening before a big test can work against you. Put it this way–a marathoner wouldn't go out for a sprint before the day of a big race. Mental marathoners–such as yourself–should not study for any more than one hour 24 hours before a Math test. That's because your brain requires some rest to be at its best. The night before your exam, spend some time with family or friends, or read a book.
- **Avoid bright screens** - You'll have to get some good shuteye the night before your test. Bright screens (such as the ones coming from your laptop, TV, or mobile device) should be avoided altogether. Staring at such a screen will keep your brain up, making it hard to drift asleep at a reasonable hour.
- **Make sure your dinner is healthy** - The meal that you have for dinner should be nutritious. Be sure to drink plenty of water as well. Load up on your complex carbohydrates, much like a marathon runner would do. Pasta, rice, and potatoes are ideal options here, as are vegetables and protein sources.
- **Get your bag ready for test day** - The night prior to your test, pack your bag with your stationery, admissions pass, ID, and any other gear that you need. Keep the bag right by your front door.
- **Make plans to reach the testing site** - Before going to sleep, ensure that you understand precisely how you will arrive at the site of the test. If parking is something you'll have to find first, plan for it. If you're dependent on public transit, then review the schedule. You should also make sure that the train/bus/subway/streetcar you use will be running. Find out about road closures as well. If a parent or friend is accompanying you, ensure that they understand what steps they have to take as well.

The Day of the Test

- **Get up reasonably early, but not too early.**
- **Have breakfast** - Breakfast improves your concentration, memory, and mood. As such, make sure the breakfast that you eat in the morning is healthy. The last thing you want to be is distracted by a grumbling tummy. If it's not your own stomach making those noises, another test taker close to you might be instead. Prevent discomfort or embarrassment by consuming a healthy breakfast. Bring a snack with you if you think you'll need it.
- **Follow your daily routine** - Do you watch Good Morning America each morning while getting ready for the day? Don't break your usual habits on the day of the test. Likewise, if coffee isn't something you drink in the morning, then don't take up the habit hours before your test. Routine consistency lets you concentrate on the main objective–doing the best you can on your test.
- **Wear layers** - Dress yourself up in comfortable layers. You should be ready for any kind of internal temperature. If it gets too warm during the test, take a layer off.
- **Get there on time** - The last thing you want to do is get to the test site late. Rather, you should be there 45 minutes prior to the start of the test. Upon your arrival, try not to hang out with anybody who is nervous. Any anxious energy they exhibit shouldn't influence you.
- **Leave the books at home** - No books should be brought to the test site. If you start developing anxiety before the test, books could encourage you to do some last-minute studying, which will only hinder you. Keep the books far away–better yet, leave them at home.
- **Make your voice heard** - If something is off, speak to a proctor. If medical attention is needed or if you'll require anything, consult the proctor prior to the start of the test. Any doubts you have should be clarified. You should be entering the test site with a state of mind that is completely clear.

- **Have faith in yourself** - When you feel confident, you will be able to perform at your best. When you are waiting for the test to begin, envision yourself receiving an outstanding result. Try to see yourself as someone who knows all the answers, no matter what the questions are. A lot of athletes tend to use this technique– particularly before a big competition. Your expectations will be reflected by your performance.

During your test..

- **Be calm and breathe deeply** - You need to relax before the test, and some deep breathing will go a long way to help you do that. Be confident and calm. You got this. Everybody feels a little stressed out just before an evaluation of any kind is set to begin. Learn some effective breathing exercises. Spend a minute meditating before the test starts. Filter out any negative thoughts you have. Exhibit confidence when having such thoughts.

- **Concentrate on the test** - Refrain from comparing yourself to anyone else. You shouldn't be distracted by the people near you or random noise. Concentrate exclusively on the test. If you find yourself irritated by surrounding noises, earplugs can be used to block sounds off close to you. Don't forget–the test is going to last several hours if you're taking more than one subject of the test. Some of that time will be dedicated to brief sections. Concentrate on the specific section you are working on during a particular moment. Do not let your mind wander off to upcoming or previous sections.

- **Skip challenging questions** - Optimize your time when taking the test. Lingering on a single question for too long will work against you. If you don't know what the answer is to a certain question, use your best guess, and mark the question so you can review it later on. There is no need to spend time attempting to solve something you aren't sure about. That time would be better served handling the questions you can actually answer well. You will not be penalized for getting the wrong answer on a test like this.

- **Try to answer each question individually** - Focus only on the question you are working on. Use one of the test-taking strategies to solve the problem. If you aren't able to come up with an answer, don't get frustrated. Simply skip that question, then move onto the next one.

- **Don't forget to breathe!** Whenever you notice your mind wandering, your stress levels boosting, or frustration brewing, take a thirty-second break. Shut your eyes, drop your pencil, breathe deeply, and let your shoulders relax. You will end up being more productive when you allow yourself to relax for a moment.
- **Optimize your breaks** - When break time comes, use the restroom, have a snack, and reactivate your energy for the subsequent section. Doing some stretches can help stimulate your blood flow.

After your test..

- **Take it easy** - You will need to set some time aside to relax and decompress once the test has concluded. There is no need to stress yourself out about what you could've said, or what you may have done wrong. At this point, there's nothing you can do about it. Your energy and time would be better spent on something that will bring you happiness for the remainder of your day.

Time to Test

Time to refine your skill with a practice examination

Take practice GACE Elementary Education Math Tests to simulate the test day experience. After you've finished, score your tests using the answer keys.

Before You Start

- You'll need a pencil and a timer to take the test.
- It's okay to guess. There is no penalty for wrong answers.
- Use the answer sheet provided to record your answers.
- The GACE Math test contains a formula sheet, which displays formulas relating to geometric measurement and certain algebra concepts. Formulas are provided to test- takers so that they may focus on application, rather than the memorization, of formulas.
- After you've finished the test, review the answer key to see where you went wrong.

Good luck!

GACE Math

Practice Test 1

2024

Total number of questions: 40

Total time: 65 Minutes

Calculators are not allowed for this test.

GACE Math Practice Test Answer Sheet

Remove (or photocopy) this answer sheet and use it to complete the practice test.

GACE Math Practice Test 1

1	Ⓐ Ⓑ Ⓒ Ⓓ	16	Ⓐ Ⓑ Ⓒ Ⓓ	31	Ⓐ Ⓑ Ⓒ Ⓓ
2	Ⓐ Ⓑ Ⓒ Ⓓ	17	Ⓐ Ⓑ Ⓒ Ⓓ	32	Ⓐ Ⓑ Ⓒ Ⓓ
3	Ⓐ Ⓑ Ⓒ Ⓓ	18	Ⓐ Ⓑ Ⓒ Ⓓ	33	Ⓐ Ⓑ Ⓒ Ⓓ
4	Ⓐ Ⓑ Ⓒ Ⓓ	19	Ⓐ Ⓑ Ⓒ Ⓓ	34	Ⓐ Ⓑ Ⓒ Ⓓ
5	Ⓐ Ⓑ Ⓒ Ⓓ	20	Ⓐ Ⓑ Ⓒ Ⓓ	35	Ⓐ Ⓑ Ⓒ Ⓓ
6	Ⓐ Ⓑ Ⓒ Ⓓ	21	Ⓐ Ⓑ Ⓒ Ⓓ	36	Ⓐ Ⓑ Ⓒ Ⓓ
7	Ⓐ Ⓑ Ⓒ Ⓓ	22	Ⓐ Ⓑ Ⓒ Ⓓ	37	Ⓐ Ⓑ Ⓒ Ⓓ
8	Ⓐ Ⓑ Ⓒ Ⓓ	23	Ⓐ Ⓑ Ⓒ Ⓓ	38	Ⓐ Ⓑ Ⓒ Ⓓ
9	Ⓐ Ⓑ Ⓒ Ⓓ	24	Ⓐ Ⓑ Ⓒ Ⓓ	39	Ⓐ Ⓑ Ⓒ Ⓓ
10	Ⓐ Ⓑ Ⓒ Ⓓ	25	Ⓐ Ⓑ Ⓒ Ⓓ	40	Ⓐ Ⓑ Ⓒ Ⓓ
11	Ⓐ Ⓑ Ⓒ Ⓓ	26	Ⓐ Ⓑ Ⓒ Ⓓ		
12	Ⓐ Ⓑ Ⓒ Ⓓ	27	Ⓐ Ⓑ Ⓒ Ⓓ		
13	Ⓐ Ⓑ Ⓒ Ⓓ	28	Ⓐ Ⓑ Ⓒ Ⓓ		
14	Ⓐ Ⓑ Ⓒ Ⓓ	29	Ⓐ Ⓑ Ⓒ Ⓓ		
15	Ⓐ Ⓑ Ⓒ Ⓓ	30	Ⓐ Ⓑ Ⓒ Ⓓ		

Formula Sheet

Perimeter / Circumference

Rectangle

$Perimeter = 2(length) + 2(width)$

Circle

$Circumference = 2\pi(radius)$

Area

Circle

$Area = \pi(radius)^2$

Triangle

$Area = \frac{1}{2}(base)(height)$

Parallelogram

$Area = (base)(height)$

Trapezoid

$Area = \frac{1}{2}(base_1 + base_2)(height)$

Volume

Prism/Cylinder

$Volume = (area\ of\ the\ base)(height)$

Pyramid/Cone

$Volume = \frac{1}{3}(area\ of\ the\ base)(height)$

Sphere

$Volume = \frac{4}{3}\pi(radius)^3$

Length

1 foot = 12 inches

1 yard = 3 feet

1 mile = 5,280 feet

1 meter = 1,000 millimeters

1 meter = 100 centimeters

1 kilometer = 1,000 meters

1 mile ≈ 1.6 kilometers

1 inch = 2.54 centimeters

1 foot ≈ 0.3 meter

Capacity / Volume

1 cup = 8 fluid ounces

1 pint = 2 cups

1 quart = 2 pints

1 gallon = 4 quarts

1 gallon = 231 cubic inches

1 liter = 1,000 milliliters

1 liter ≈ 0.264 gallon

Weight

1 pound = 16 ounces

1 ton = 2,000 pounds

1 gram = 1,000 milligrams

1 kilogram = 1,000 grams

1 kilogram ≈ 2.2 pounds

1 ounce ≈ 28.3 grams

1) The capacity of a red box is 20% bigger than the capacity of a blue box. If the red box can hold 30 equal sized books, how many of the same books can the blue box hold?

   A. 9
   B. 15
   C. 21
   D. 25
   E. 30

2) Kim spent $35 for pants. This was $10 less than triple what she spent for a shirt. How much was the shirt?

   A. $11
   B. $13
   C. $15
   D. $17
   E. $21

3) What is the greatest integer less than $-\frac{32}{5}$?

   A. 0
   B. -2
   C. -4
   D. -6
   E. -7

4) The measure of the angles of a triangle are in the ratio $1:3:5$. What is the measure of the largest angle?

   A. 20°
   B. 45°
   C. 85°
   D. 100°
   E. 180°

5) In the figure below, line A is parallel to line B. what is the value of x?

   A. 28
   B. 46
   C. 50
   D. 55
   E. 65

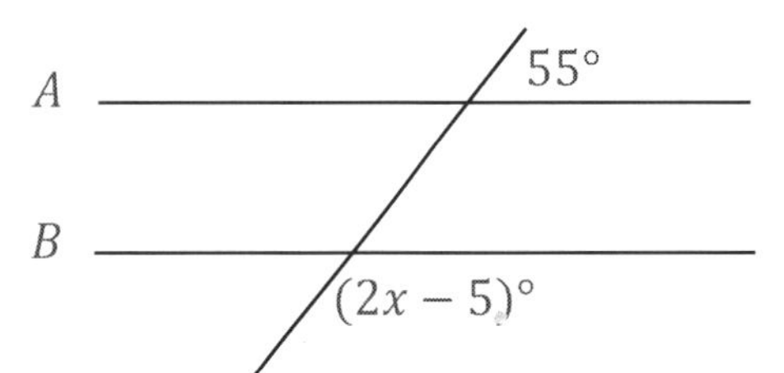

6) In the infinitely repeating decimal below, 1 is the first digit in the repeating pattern. What is the $68th$ digit? $\frac{1}{7} = 0.\overline{142857}$

A. 1
B. 2
C. 4
D. 5
E. 7

7) In the following figure, $ABCD$ is a rectangle. If $a = \sqrt{3}$, and $b = 2a$, find the area of the shaded region. (the shaded region is a trapezoid)

A. $2\sqrt{3}$
B. $3\sqrt{3}$
C. $4\sqrt{3}$
D. $6\sqrt{3}$
E. $8\sqrt{3}$

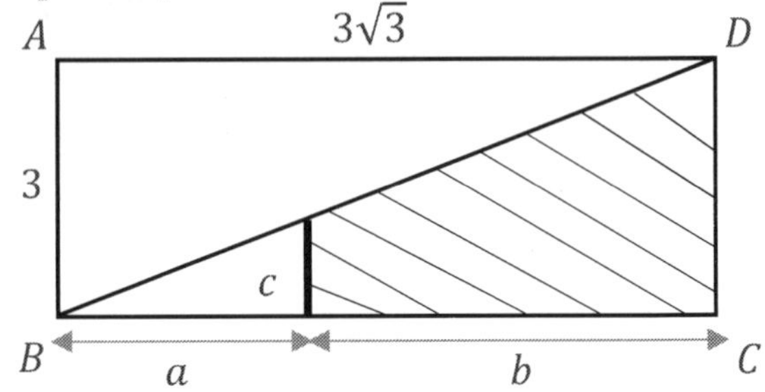

8) The supplement angle of a 45° angle is:

A. 135°
B. 105°
C. 90°
D. 35°
E. 15°

9) Anna opened an account with a deposit of $3,000. This account earns 5% simple interest annually. How many years will it take her to earn $600 on her $3,000 deposit?

A. 2
B. 4
C. 5
D. 6
E. 8

10) A list of consecutive integers begins with k and ends with n. If $n - k = 46$, how many integers are in the list?

A. 23
B. 38
C. 46
D. 47
E. 58

11) Tom picked $2\frac{2}{5}$ baskets of apples, and Sam picked $1\frac{3}{4}$ baskets of apples. How many baskets total did they pick?

A. $1\frac{2}{3}$
B. $2\frac{1}{12}$
C. $3\frac{20}{23}$
D. $4\frac{3}{20}$
E. $5\frac{1}{12}$

12) A piece of paper that is $2\frac{3}{5}$ feet long is cut into 2 pieces of different lengths. The shorter piece has a length of x feet. Which inequality expresses all possible values of x?

A. $x < 2\frac{1}{10}$
B. $x > 2$
C. $x < 2\frac{3}{5}$
D. $x > 1\frac{3}{10}$
E. $x < 1\frac{3}{10}$

13) In an academy, course grades range from 0 to 100. Anna took 5 courses and her mean course grade was 80. William took 8 courses. If both students have the same sum of course grades, what was William's mean?

A. 50
B. 65
C. 70
D. 80
E. 85

14) The set S consists of all odd numbers greater than 5 and less than 30. What is the mean of the numbers in S?

A. 11
B. 13
C. 17
D. 18
E. 23

15)In a group of 45 student, 60% can't swim. How many students can swim?

A. 13
B. 18
C. 22
D. 23
E. 35

16)$8,400 are distributed equally among 14 person. How much money will each person get?

A. $400
B. $450
C. $584
D. $600
E. $800

17)A box contains 6 green sticks, 4 blue sticks, and 2 yellow sticks. Emma picks one without looking. What is the probability that the stick will be green?

A. $\frac{1}{2}$
B. $\frac{1}{3}$
C. $\frac{1}{4}$
D. $\frac{2}{5}$
E. $\frac{3}{2}$

18)The price of a Chocolate was raised from $5.40 to $5.67. What was the percent increase in the price?

A. 4%
B. 5%
C. 6%
D. 8%
E. 10%

19)In a box of blue and black marbles, the ratio of blue marbles to black marbles is $4:3$. If the box contains 150 black marbles, how many blue marbles are there?

A. 100
B. 150
C. 200
D. 300
E. 600

20) $\frac{5}{8}$ of a number is 90. Find the number.

A. 144
B. 270
C. 450
D. 720
E. 800

21) A juice mixture contains $\frac{5}{14}$ jar of cherry juice and $\frac{5}{70}$ jar of apple juice. How many jars of cherry juice per jar of apple juice does the mixture contain?

A. 70
B. 14
C. 10
D. 7
E. 5

22) The set of possible values of n is $\{5, 3, 7\}$. What is the set of possible values of m if $2m = n + 5$?

A. $\{2, 4, 7\}$
B. $\{3, 2, 5\}$
C. $\{4, 5, 8\}$
D. $\{5, 4, 6\}$
E. $\{6, 5, 8\}$

23) If $x = 25$, then which of the following equations are correct?

A. $x + 10 = 40$
B. $4x = 100$
C. $3x = 70$
D. $\frac{x}{2} = 12$
E. $\frac{x}{3} = 8$

24) Jack scored a mean of 80 per test in his first 4 tests. In his 5th test, he scored 90. What was Jack's mean score for the 5 tests?

A. 70
B. 75
C. 80
D. 82
E. 93

25) The volume of a cube is less than $64\ m^3$. Which of the following can be the cube's side?

A. $2\ m$
B. $4\ m$
C. $8\ m$
D. $10\ m$
E. $11\ m$

26) What is the area of an isosceles right triangle that has one leg that measures $6\ cm$?

A. $16\ cm^2$
B. $18\ cm^2$
C. $24\ cm^2$
D. $32\ cm^2$
E. $36\ cm^2$

27) If $0.00104 = \frac{104}{x}$, what is the value of x?

A. 1,000
B. 10,000
C. 100,000
D. 1,000,000
E. 10,000,000

28) A bag is filled with numbered cards from 1 to 15 and picked on at random. What is the probability that the card picked is number 8?

A. $\frac{8}{15}$
B. $\frac{7}{15}$
C. $\frac{5}{15}$
D. $\frac{2}{15}$
E. $\frac{1}{15}$

29) What is the value of x in the figure below?

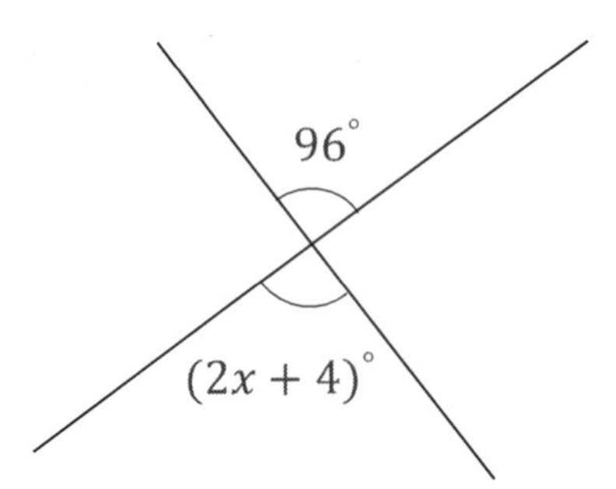

A. 21
B. 26
C. 36
D. 46
E. 48

30) How many different two-digit numbers can be formed from the digits 6, 7, and 5, if the numbers must be even and no digit can be repeated?

A. 1
B. 2
C. 3
D. 4
E. 5

31) A rectangular concrete driveway is 25 feet long, 6 feet wide, and 24 inches thick. What is the volume of the concrete?

A. 300 ft^3
B. 600 ft^3
C. 660 ft^3
D. 963 ft^3
E. 1,800 ft^3

32) $200(3+0.01)^2 - 200 =$

A. 201.55
B. 361.08
C. 702.88
D. 1,612.02
E. 1,812.02

33) If 360 kg of vegetables is packed in 90 boxes, how much vegetables will each box contain?

A. 2.5 kg
B. 3 kg
C. 4 kg
D. 6.5 kg
E. 7 kg

34) Each number in a sequence is 4 more than twice the number that comes just before it. If 84 is a number in the sequence, what number comes just before it?

A. 26
B. 35
C. 40
D. 52
E. 88

35) $[6 \times (-24) + 8] - (-4) + [4 \times 5] \div 2 = ?$

A. 148
B. 132
C. -122
D. -136
E. -144

36) A rectangle has 14 cm wide and 5 cm length. What is the perimeter of this rectangle?

A. 19 cm
B. 28 cm
C. 33 cm
D. 38 cm
E. 41 cm

37) What is the value of the following expression? $3\frac{1}{4} + 2\frac{4}{16} + 1\frac{3}{8} + 5\frac{1}{2}$

A. $3\frac{10}{14}$
B. $4\frac{1}{2}$
C. $12\frac{4}{16}$
D. $12\frac{3}{8}$
E. $12\frac{4}{8}$

38) A certain insect has a mass of 85 milligrams. What is the insect's mass in grams?

A. 0.085
B. 0.08
C. 0.85
D. 8.5
E. 85

39) Removing which of the following numbers will change the average of the numbers to 6?

$$1, 4, 5, 8, 11, 12$$

A. 1
B. 4
C. 5
D. 8
E. 11

40) If $m = 6$ and $n = -3$, what is the value of $\frac{5-9(3+n)}{3m-5(2-n)}$ =?

A. $\frac{2}{7}$
B. $\frac{3}{7}$
C. $-\frac{4}{7}$
D. $\frac{5}{7}$
E. $-\frac{5}{7}$

End of CBEST Math Practice Test 1

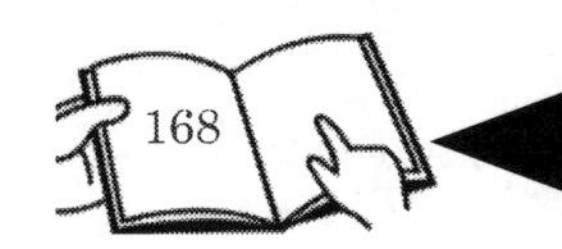

GACE Math

Practice Test 2

2024

Total number of questions: 40

Total time: 65 Minutes

Calculators are not allowed for this test.

GACE Math Practice Test Answer Sheet

Remove (or photocopy) this answer sheet and use it to complete the practice test.

GACE Math Practice Test 2					
1	Ⓐ Ⓑ Ⓒ Ⓓ	16	Ⓐ Ⓑ Ⓒ Ⓓ	31	Ⓐ Ⓑ Ⓒ Ⓓ
2	Ⓐ Ⓑ Ⓒ Ⓓ	17	Ⓐ Ⓑ Ⓒ Ⓓ	32	Ⓐ Ⓑ Ⓒ Ⓓ
3	Ⓐ Ⓑ Ⓒ Ⓓ	18	Ⓐ Ⓑ Ⓒ Ⓓ	33	Ⓐ Ⓑ Ⓒ Ⓓ
4	Ⓐ Ⓑ Ⓒ Ⓓ	19	Ⓐ Ⓑ Ⓒ Ⓓ	34	Ⓐ Ⓑ Ⓒ Ⓓ
5	Ⓐ Ⓑ Ⓒ Ⓓ	20	Ⓐ Ⓑ Ⓒ Ⓓ	35	Ⓐ Ⓑ Ⓒ Ⓓ
6	Ⓐ Ⓑ Ⓒ Ⓓ	21	Ⓐ Ⓑ Ⓒ Ⓓ	36	Ⓐ Ⓑ Ⓒ Ⓓ
7	Ⓐ Ⓑ Ⓒ Ⓓ	22	Ⓐ Ⓑ Ⓒ Ⓓ	37	Ⓐ Ⓑ Ⓒ Ⓓ
8	Ⓐ Ⓑ Ⓒ Ⓓ	23	Ⓐ Ⓑ Ⓒ Ⓓ	38	Ⓐ Ⓑ Ⓒ Ⓓ
9	Ⓐ Ⓑ Ⓒ Ⓓ	24	Ⓐ Ⓑ Ⓒ Ⓓ	39	Ⓐ Ⓑ Ⓒ Ⓓ
10	Ⓐ Ⓑ Ⓒ Ⓓ	25	Ⓐ Ⓑ Ⓒ Ⓓ	40	Ⓐ Ⓑ Ⓒ Ⓓ
11	Ⓐ Ⓑ Ⓒ Ⓓ	26	Ⓐ Ⓑ Ⓒ Ⓓ		
12	Ⓐ Ⓑ Ⓒ Ⓓ	27	Ⓐ Ⓑ Ⓒ Ⓓ		
13	Ⓐ Ⓑ Ⓒ Ⓓ	28	Ⓐ Ⓑ Ⓒ Ⓓ		
14	Ⓐ Ⓑ Ⓒ Ⓓ	29	Ⓐ Ⓑ Ⓒ Ⓓ		
15	Ⓐ Ⓑ Ⓒ Ⓓ	30	Ⓐ Ⓑ Ⓒ Ⓓ		

Formula Sheet

Perimeter / Circumference

Rectangle

$Perimeter = 2(length) + 2(width)$

Circle

$Circumference = 2\pi(radius)$

Area

Circle

$Area = \pi(radius)^2$

Triangle

$Area = \frac{1}{2}(base)(height)$

Parallelogram

$Area = (base)(height)$

Trapezoid

$Area = \frac{1}{2}(base_1 + base_2)(height)$

Volume

Prism/Cylinder

$Volume = (area\ of\ the\ base)(height)$

Pyramid/Cone

$Volume = \frac{1}{3}(area\ of\ the\ base)(height)$

Sphere

$Volume = \frac{4}{3}\pi(radius)^3$

Length

1 foot = 12 inches

1 yard = 3 feet

1 mile = 5,280 feet

1 meter = 1,000 millimeters

1 meter = 100 centimeters

1 kilometer = 1,000 meters

1 mile ≈ 1.6 kilometers

1 inch = 2.54 centimeters

1 foot ≈ 0.3 meter

Capacity / Volume

1 cup = 8 fluid ounces

1 pint = 2 cups

1 quart = 2 pints

1 gallon = 4 quarts

1 gallon = 231 cubic inches

1 liter = 1,000 milliliters

1 liter ≈ 0.264 gallon

Weight

1 pound = 16 ounces

1 ton = 2,000 pounds

1 gram = 1,000 milligrams

1 kilogram = 1,000 grams

1 kilogram ≈ 2.2 pounds

1 ounce ≈ 28.3 grams

1) A shoe originally priced at $45.00 was on sale for 15% off. Nick received a 20% employee discount applied to the sale price. How much did Nick pay for the shoes?

A. $30.60
B. $34.50
C. $37.30
D. $38.25
E. $42.25

2) Which of the following values when entered in the box will satisfy the statement below?

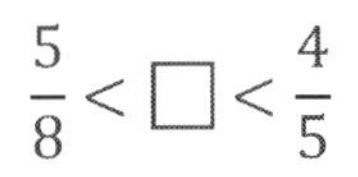

$$\frac{5}{8} < \square < \frac{4}{5}$$

A. $\frac{2}{5}$
B. $\frac{3}{4}$
C. $\frac{6}{10}$
D. $\frac{3}{5}$
E. $\frac{7}{8}$

3) What is the probability a D on of spinning the spinner?

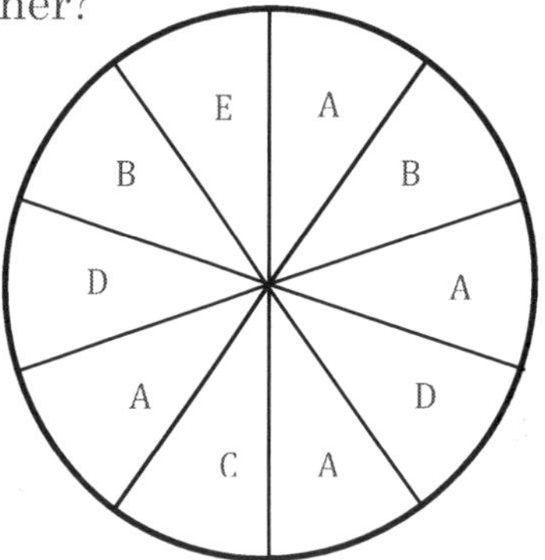

A. 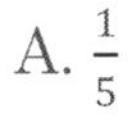 $\frac{1}{5}$
B. $\frac{1}{10}$
C. $\frac{3}{10}$
D. $\frac{2}{5}$
E. $\frac{3}{10}$

4) Which of the following is a factor of 45?

A. 7
B. 9
C. 11
D. 13
E. 14

5) By what percent did the price of a shirt increase if its price was increased from \$15.30 to \$18.36?

A. 10%
B. 12%
C. 16%
D. 20%
E. 22%

6) The greatest common factor of 32 and x is 8. How many possible values for x are greater than 10 and less than 60?

A. 1
B. 4
C. 6
D. 7
E. 9

7) A box contains 6 strawberry candies, 4 orange candies, and 3 banana candies. If Roberto selects 2 candies at random from this box, without replacement, what is the probability that both candies are not orange?

A. $\frac{1}{28}$
B. $\frac{2}{13}$
C. $\frac{6}{13}$
D. $\frac{1}{3}$
E. $\frac{2}{3}$

8) How many integers are between $\frac{7}{2}$ and $\frac{30}{4}$?

A. 3
B. 4
C. 6
D. 10
E. 12

9) In a certain state, the sales tax rate increased from 8% to 8.5%. What was the increase in the sales tax on a \$250 item?

A. \$0.5
B. \$1.00
C. \$1.25
D. \$1.90
E. \$2.30

10) Triangle ABC is graphed on a coordinate grid with vertices at $A\ (-3,-2)$, $B\ (-1,4)$ and $C\ (7,9)$. Triangle ABC is reflected over x axes to create triangle $A'\ B'\ C'$. Which order pair represents the coordinate of C'?

A. $(-7,-9)$
B. $(-7,9)$
C. $(7,-9)$
D. $(7,9)$
E. (9,7)

11) Made a list of all possible products of 2 different numbers in the set below. What fraction of the products are odd?

$$\{1,4,6,5,7\}$$

A. $\frac{2}{5}$
B. $\frac{3}{10}$
C. $\frac{7}{10}$
D. $\frac{4}{15}$
E. $\frac{8}{17}$

12) If $5n$ is a positive even number, how many odd numbers are in the range from $5n$ up to and including $5n + 6$?

A. 1
B. 2
C. 3
D. 4
E. 5

13) If the actual Height of the building is 2,760 centimeters, then what is the scale of the diagram of the building?

A. $1\ unit = 552\ centimeter$
B. $1\ unit = 650\ centimeter$
C. $1\ unit = 680\ centimeter$
D. $1\ unit = 690\ centimeter$
E. $1\ unit = 700\ centimeter$

14) If $b = 2$ and $\frac{a}{4} = b$, what is the value of $a^2 + 4b$?

A. 54
B. 66
C. 72
D. 76
E. 81

15) Which percentage is closest in value to 0.0099?

A. 0.1%
B. 1%
C. 2%
D. 9%
E. 100%

16) What is the surface area of the cylinder below?

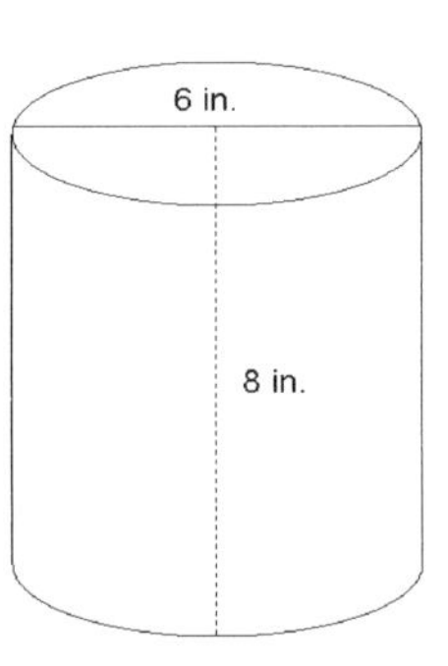

A. $48\pi\ in^2$
B. $57\pi\ in^2$
C. $66\pi\ in^2$
D. $288\pi\ in^2$
E. $400\pi\ in^2$

17) A train travels 1,500 miles from New York to Oklahoma. The train covers the first 280 miles in 4 hours. If the train continues to travel at this rate, how many more hours will it take to reach Oklahoma City? Round your answer to the nearest whole hour.

A. 12
B. 15
C. 17
D. 20
E. 22

18) The sales price of a laptop is $1,912.50, which is 15% off the original price. What is the original price of the laptop?

A. $2,750
B. $2,250
C. $1,625.625
D. $956.25
E. $286.875

19) In a scale diagram, 0.15 inch represents 150 feet. How many inches represent 2.5 *feet*?

A. 0.001 *in*
B. 0.002 *in*
C. 0.0025 *in*
D. 0.01 *in*
E. 0.012 *in*

20) If $\frac{3}{7}$ of Z is 54, what is $\frac{2}{5}$ of Z?

A. 44.2
B. 46.3
C. 48.4
D. 50.4
E. 60.6

21) A car travels at a speed of 72 miles per hour. How far will it travel in 8 hours?

A. 576
B. 540
C. 480
D. 432
E. 272

22) If Sam spent \$60 on sweets and he spent 25% of the selling price for the tip, how much did he spend?

A. \$66
B. \$69
C. \$72
D. \$75
E. \$77

23) Which of the following numbers has factors that include the smallest factor (other than 1) of 95?

A. 25
B. 28
C. 32
D. 39
E. 45

24) $\frac{4^2+3^2+(-5)^2}{(9+10-11)^2}=?$

A. $\frac{25}{32}$
B. $-\frac{25}{32}$
C. 56
D. -56
E. 64

25) Angle A and angle B are supplementary. The measure of angle A is 2 times the measure of angle B. What is the measure of angle A in degrees?

A. 100°
B. 120°
C. 140°
D. 160°
E. 170°

26) Tomas is 6 feet 8.5 inches tall, and Alex is 5 feet 3 inches tall. What is the difference in height, in inches, between Alex and Tomas?

A. 2.5
B. 7.5
C. 12.5
D. 17.5
E. 19.5

27) Yesterday Kylie writes 10% of her homework. Today she writes another 18% of the entire homework. What fraction of the homework is left for her to write?

A. $\frac{4}{25}$
B. $\frac{7}{25}$
C. $\frac{10}{25}$
D. $\frac{18}{25}$
E. $\frac{21}{25}$

28) In a box of blue and yellow pens, the ratio of yellow pens to blue pens is $2:3$. If the box contains 9 blue pens, how many yellow pens are there?

A. 2
B. 3
C. 4
D. 5
E. 6

29) What decimal is equivalent to $-\frac{6}{9}$?

A. $-0.\overline{5}$
B. $-0.\overline{6}$
C. $-0.\overline{65}$
D. $-0.\overline{7}$
E. $-0.\overline{75}$

30) The area of a circle is 81π. What is the diameter of the circle?

A. 3
B. 6
C. 8
D. 9
E. 18

31) Five years ago, Amy was three times as old as Mike was. If Mike is 10 years old now, how old is Amy?

A. 4
B. 8
C. 12
D. 14
E. 20

32) How many positive even factors of 68 are greater than 26 and less than 60?

A. 0
B. 1
C. 2
D. 4
E. 6

33) The ratio of two sides of a parallelogram is $2:3$. If its perimeter is $40\ cm$, find the length of its sides.

A. $6\ cm, 12\ cm$
B. $8\ cm, 12\ cm$
C. $10\ cm, 14\ cm$
D. $12\ cm, 16\ cm$
E. $14\ cm, 18\ cm$

34) What is the value of x in the following equation? $\frac{3}{4}(x-2) = 3\left(\frac{1}{6}x - \frac{3}{2}\right)$

A. $\frac{1}{4}$
B. $-\frac{3}{4}$
C. -3
D. 6
E. -12

35) If x can be any integer, what is the greatest possible value of the expression $2 - x^2$?

A. -1
B. 0
C. 2
D. 3
E. 4

36) A store has a container of handballs: 6 green, 5 blue, 8 white, and 10 yellow. If one ball is picked from the container at random, what is the probability that it will be green?

A. $\frac{1}{5}$
B. $\frac{6}{11}$
C. $\frac{6}{29}$
D. $\frac{8}{25}$
E. $\frac{11}{25}$

37) Emma answered 9 out of 45 questions on a test incorrectly. What percentage of the questions did she answer correctly?

A. 10%
B. 40%
C. 68%
D. 80%
E. 92%

38) If 30% of a number is 12, what is the number?

A. 12
B. 25
C. 40
D. 45
E. 50

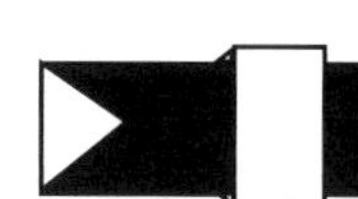

39) If Anna multiplies her age by 5 and then adds 3, she will get a number equal to her mother's age. If x is her mother's age, what is Anna's age in terms of x?

A. $\frac{x-3}{5}$
B. $\frac{x-5}{3}$
C. $3x + 5$
D. $5x - 3$
E. $x - 3$

40) A line connects the midpoint of AB (point E), with point C in the square $ABCD$. Calculate the area of the acquired trapezoid shape if the square has a side of $4\ m$.

A. $4\ cm^2$
B. $12\ cm^2$
C. $15\ cm^2$
D. $18\ cm^2$
E. $24\ cm^2$

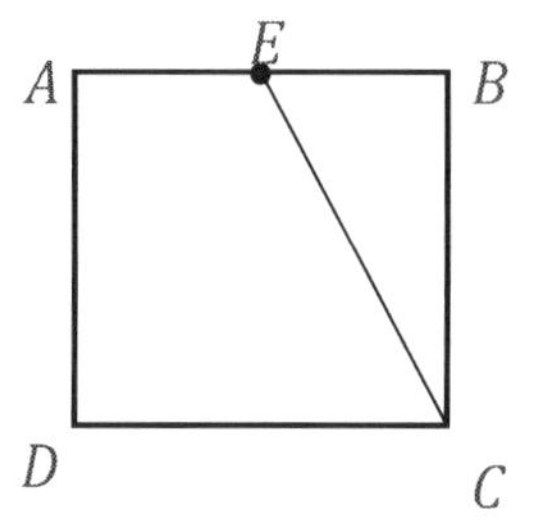

End of GACE Math Practice Test 2

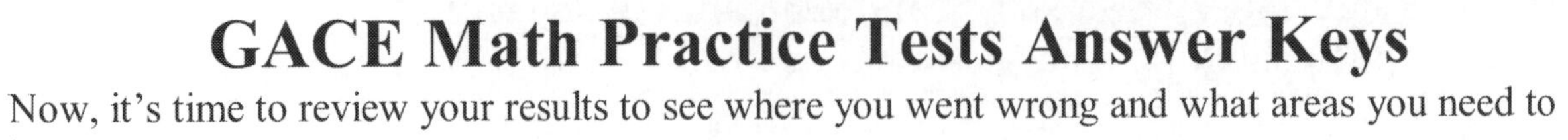

GACE Math Practice Tests Answer Keys

Now, it's time to review your results to see where you went wrong and what areas you need to improve.

GACE Practice Test 1				GACE Practice Test 2			
1	D	**21**	E	**1**	A	**21**	A
2	C	**22**	D	**2**	B	**22**	D
3	E	**23**	B	**3**	A	**23**	E
4	D	**24**	D	**4**	B	**24**	A
5	E	**25**	A	**5**	D	**25**	B
6	C	**26**	B	**6**	B	**26**	D
7	C	**27**	C	**7**	C	**27**	D
8	A	**28**	E	**8**	B	**28**	E
9	B	**29**	D	**9**	C	**29**	B
10	D	**30**	B	**10**	C	**30**	E
11	D	**31**	A	**11**	B	**31**	E
12	E	**32**	D	**12**	C	**32**	B
13	A	**33**	C	**13**	B	**33**	B
14	D	**34**	C	**14**	C	**34**	E
15	B	**35**	C	**15**	A	**35**	C
16	D	**36**	D	**16**	C	**36**	C
17	A	**37**	D	**17**	C	**37**	D
18	B	**38**	A	**18**	B	**38**	C
19	C	**39**	E	**19**	C	**39**	A
20	A	**40**	E	**20**	D	**40**	B

GACE Math Practice Tests Answers and Explanations

GACE Math Practice Test 1

1) Choice D is correct

The capacity of a red box is 20% bigger than the capacity of a blue box and it can hold 30 books. Therefore, we want to find a number that 20% bigger than that number is 30. Let x be that number. Then: $1.20 \times x = 30$. Divide both sides of the equation by 1.2. Then: $x = \frac{30}{1.20} = 25$

2) Choice C is correct

Convert everything into an equation: $35 = (3 \times \text{shirt}) - 10$

Now, solve the equation: $45 = 3 \text{ shirt} \rightarrow \text{shirt} = \frac{45}{3} = 15$. The price of the shirt was $15.

3) Choice E is correct

First, convert the improper fraction to a mixed number: $-\frac{32}{5} = -6\frac{2}{5}$

The two closest integers to this fraction are -7 and -6.

The integer less than $-\frac{32}{5}$ is -7.

4) Choice D is correct

Let x equal the smallest angle of the triangle. Then, the three angles are $x, 3x$, and $5x$. The sum of the angles of a triangle is 180. Set up an equation using this to find x:

$x + 3x + 5x = 180 \rightarrow 9x = 180 \rightarrow x = 20$

Since the question asks for the measure of the largest angle, $5x = 5(20) = 100^\circ$

5) Choice E is correct

The angle $(2x - 5)$ and 55 are supplementary angles. Therefore:

$(2x - 5) + 55 = 180 \rightarrow 2x + 50 = 180 \rightarrow 2x = 180 - 50 \rightarrow 2x = 130 \rightarrow x = \frac{130}{2} \rightarrow x = 65$

6) Choice C is correct

There are 6 digits in the repeating decimal (0.142857), so digit 1 would be the first, seventh, thirteenth digit and so on. To find the 68th digit, divide 68 by 6.

$68 \div 6 = 11r2$

7) Choice C is correct

Based on triangle similarity theorem: $\frac{a}{a+b} = \frac{c}{3} \rightarrow c = \frac{3a}{a+b} = \frac{3\sqrt{3}}{3\sqrt{3}} = 1 \rightarrow$ Area of shaded region is: $\left(\frac{c+3}{2}\right)(b) = \frac{4}{2} \times 2\sqrt{3} = 4\sqrt{3}$

8) Choice A is correct

Two Angles are supplementary when they add up to 180 degrees.

$135^\circ + 45^\circ = 180^\circ$

9) Choice B is correct

Use simple interest formula:

$I = prt$ $(I = interest, p = principal, r = rate, t = time)$

$I = prt \rightarrow 600 = (3{,}000)(0.05)(t) \rightarrow 600 = 150t \rightarrow t = 4$

10) Choice D is correct

Consider the case where $k = 1$

$n - k = 46 \rightarrow n - 1 = 46 \rightarrow n - 1 + 1 = 46 + 1 \rightarrow n = 47$

The list of integers from 1 to 47 contains 47 numbers.

11) Choice D is correct

To solve, add the two given fractions: $2\frac{2}{5} + 1\frac{3}{4}$

The common denominator is 20: $2\frac{8}{20} + 1\frac{15}{20} = 3\frac{23}{20} = 4\frac{3}{20}$

12) Choice E is correct

The original piece of paper is $2\frac{3}{5}$ feet long.

The shorter piece is x feet long, and it must be less than half the length of the original piece of paper. Since half of $2\frac{3}{5}$ is $1\frac{3}{10}$ it follows that $x < 1\frac{3}{10}$.

13) Choice A is correct

First, find the sum of course grade of Anna, $average = \frac{sum\ of\ terms}{number\ of\ terms} \Rightarrow$

$80 = \frac{sum\ of\ course\ grade}{5} \rightarrow the\ sum\ of\ course\ grade = 80 \times 5 = 400$

Anna and William have the same sum of course grade, now find the Williams mean

$$average = \frac{sum\ of\ course\ grade}{number\ of\ course} \Rightarrow \frac{400}{8} = 50$$

14) Choice D is correct

List in order the odd numbers between 5 to 30: $7, 9, 11, 13, 15, 17, 19, 21, 23, 25, 27$, and 29. Since, the numbers are consecutive odd numbers, the mean and the median are equal. The median is the number in the middle. Since we have 12 numbers, the median is the average of numbers 6 and 7 which are 17 and 19. The mean (or the median) is: Mean $= \frac{17+19}{2} = 18$

15) Choice B is correct

60% of students can't swim→ $100 - 60 = 40\%$ can swim.

Then: $0.40 \times 45 = 18$

16) Choice D is correct

Money received by 14 person = \$8,400. So, the money received by one person is:

$\frac{\$8,400}{14} = \$600$

17) Choice A is correct

There are 12 sticks in the box $(6 + 4 + 2)$. So, the probability that Emma picks a green stick is: $Probability = \frac{6}{12} = \frac{1}{2}$

18) Choice B is correct

Use the percent increase expression to find the answer:

$\frac{new\ price - original\ price}{original\ price} = \frac{5.67-5.40}{5.40} = 0.05 = 5\%$

19) Choice C is correct

Let x be the number of blue marbles. Write the items in the ratio as a fraction:

$\frac{x}{150} = \frac{4}{3} \rightarrow 3x = 600 \rightarrow x = 200$

20) Choice A is correct

Let x be the number: $\frac{5}{8}x = 90 \rightarrow x = 90 \times \frac{8}{5} = \frac{720}{5} = 144$

21) Choice E is correct

Set up a proportion to solve: $\frac{\frac{5}{14}cherry}{\frac{5}{70}apple} = \frac{x\ cherry}{1\ apple} \rightarrow \frac{5}{14} \times \frac{70}{5} = x \rightarrow x = \frac{70}{14} = \frac{10}{2} \rightarrow x = 5$

22) Choice D is correct

$2m = n + 5 \rightarrow m = \frac{n+5}{2}$. Substitute each value of n to find the values of m:

$$m = \frac{5+5}{2} = \frac{10}{2} = 5$$

$$m = \frac{3+5}{2} = \frac{8}{2} = 4$$

$$m = \frac{7+5}{2} = \frac{12}{2} = 6$$

The set of m is $\{5,4,6\}$.

23) Choice B is correct

Plug in 25 for x in the equation.

A. $x + 10 = 40 \rightarrow 25 + 10 \neq 40$

B. $4x = 100 \rightarrow 4(25) = 100$

C. $3x = 70 \rightarrow 3(25) \neq 70$

D. $\frac{x}{2} = 12 \rightarrow \frac{25}{2} \neq 12$

E. $\frac{x}{3} = 8 \rightarrow \frac{25}{3} \neq 8$

Only choice B is correct.

24) Choice D is correct

Jack scored a mean of 80 per test. In the first 4 tests, the sum of scores is:

$80 \times 4 = 320$. Now, calculate the mean over the 5 tests: $\frac{320+90}{5} = \frac{410}{5} = 82$

25) Choice A is correct

Volume of the cube is less than 64 m^3. Use the formula of volume of cubes.

$Volume = (one\ side)^3 \Rightarrow 64 = (one\ side)^3$. Find the cube root of both sides.

$64 = (one\ side)^3 \rightarrow one\ side = \sqrt[3]{64} = 4\ m$

Then: $4 =$ one side. The side of the cube is less than 4. Only choice A is less than 4.

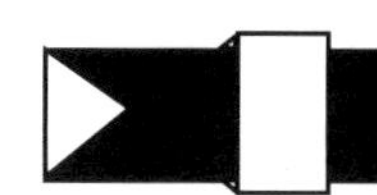

26) Choice B is correct

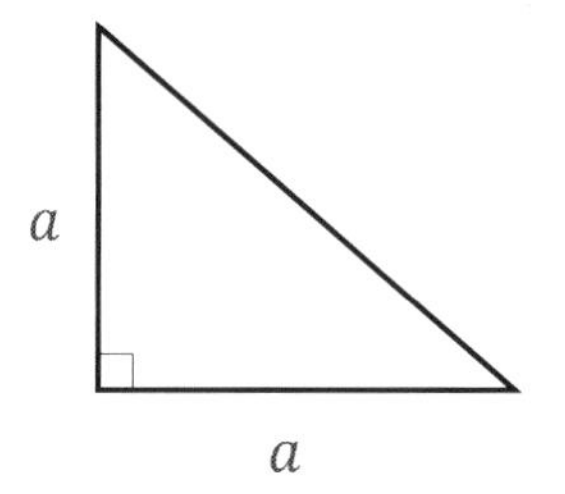

First draw an isosceles triangle. Remember that two legs of the triangle are equal.

Let put a for the legs. Then:

$a = 6 \Rightarrow$ Area of the triangle is $= \frac{1}{2}(6 \times 6) = \frac{36}{2} = 18\ cm^2$

27) Choice C is correct

Solve for x: $0.00104 = \frac{104}{x}$, multiply both sides by x, $(0.00104)(x) = \frac{104}{x}(x)$.

Simplify: $0.00104x = 104$. Divide both side by 0.00104: $\frac{0.00104x}{0.00104} = \frac{104}{0.00104}$, simplify

$x = \frac{104}{0.00104} = 100{,}000$

28) Choice E is correct

The number of cards in the bag is 15.

$Probability = \frac{number\ of\ desired\ outcomes}{number\ of\ total\ outcomes} = \frac{1}{15}$

29) Choice D is correct

$(2x + 4)^{\circ}$ and 96° are vertical angles. Vertical angles are equal in measure.

Then: $2x + 4 = 96 \rightarrow 2x = 92 \rightarrow x = 46$

30) Choice B is correct

The two-digit numbers must be even, so the only possible two-digit numbers must end in 6, since 6 is the only even digit given in the problem. Since the numbers cannot be repeated, the only possibilities for two-digit even numbers are 76 and 56. Thus, the answer is two possible two-digit numbers.

31) Choice A is correct

First convert 24 inches to feet. 12 inch = 1 feet, thus: $24 \div 12 = 2\ feet$. Then, calculate the volume, in cubic feet: $25 \times 6 \times 2 = 300\ ft^3$

32) Choice D is correct

First calculate exponents value, then multiplying and subtracting:

$200(3 + 0.01)^2 - 200 = 200(3.01)^2 - 200 = 200(9.0601) - 200 = 1{,}612.02$

33) Choice C is correct

Since 90 boxes contain 360 kg vegetable. Therefore, 1 box contains $\frac{360\ kg}{90} = 4\ kg$ vegetable.

34) Choice C is correct

Let n represent a number in the sequence, and let x represent the number that comes just before n. $n = 4 + 2x \rightarrow 84 = 4 + 2x \rightarrow 80 = 2x \rightarrow x = 40$

35) Choice C is correct

Use PEMDAS (order of operation): $[6 \times (-24) + 8] - (-4) + [4 \times 5] \div 2 =$

$[-144 + 8] - (-4) + [20] \div 2 = [-144 + 8] + 4 + 10 = [-136] + 4 + 10 = -122$

36) Choice D is correct

Perimeter of rectangle is equal to the sum of all the sides of the rectangle:

Perimeter $= 2(14) + 2(5) = 28 + 10 = 38\ cm$

37) Choice D is correct

$3\frac{1}{4} + 2\frac{4}{16} + 1\frac{3}{8} + 5\frac{1}{2}$. Convert all the fractions to a common denominator (16):

$3\frac{4}{16} + 2\frac{4}{16} + 1\frac{6}{16} + 5\frac{8}{16} = (3 + 2 + 1 + 5) + \left(\frac{4+4+6+8}{16}\right) = 11 + 1\frac{6}{16} = 12\frac{6}{16} = 12\frac{3}{8}$

38) Choice A is correct

One gram is equal to 1,000 milligrams, or 1 milligram is equal to $\frac{1}{1{,}000}$ gram.

Thus, 85 milligrams $= \frac{85}{1{,}000} = 0.085$ gram

39) Choice E is correct

Check each choice provided:

A. 1 $\quad \frac{4+5+8+11+12}{5} = \frac{40}{5} = 8$

B. 4 $\quad \frac{1+5+8+11+12}{5} = \frac{37}{5} = 7.4$

C. 5 $\quad \frac{1+4+8+11+12}{5} = \frac{36}{5} = 7.2$

D. 8 $\quad \frac{1+4+5+11+12}{5} = \frac{33}{5} = 6.6$

E. 11 $\quad \frac{1+4+5+8+12}{5} = \frac{30}{5} = 6$

40) Choice E is correct

Substitute 6 for m and -3 for n:

$\frac{5-9(3+n)}{3m-5(2-n)} = \frac{5-9(3+(-3))}{3(6)-5(2-(-3))} = \frac{5-9(0)}{18-5(5)} = \frac{5}{18-25} = \frac{5}{-7} = -\frac{5}{7}$

GACE Math Practice Test 2

1) Choice A is correct

First, find the sale price. 15% of \$45.00 is \$6.75, so the sale price is

\$45.00 – \$6.75 = \$38.25. Next, find the price after Nick's employee discount. 20% × \$38.25 = \$7.65, so, the final price of the shoes is \$38.25 – \$7.65 = \$30.60.

2) Choice B is correct

Let's compare the fractions by converting them to decimals:

A. $\frac{2}{5} = 0.4$

B. $\frac{3}{4} = 0.75$

C. $\frac{6}{10} = 0.6$

D. $\frac{3}{5} = 0.6$

E. $\frac{7}{8} = 0.875$

Only 0.75 can be entered in the box.

$\frac{5}{8} < \square < \frac{4}{5} \rightarrow 0.625 < \square < 0.8 \rightarrow 0.625 < 0.75 < 0.8$

3) Choice A is correct

The total number of possible sections in the spinner = 10

There are two sections containing Number D. Then, the probability of spinning a D is: $\frac{2}{10} = \frac{1}{5}$

4) Choice B is correct

The factors of 45 are: $\{1, 3, 5, 9, 15, 45\}$. Only choice B is correct.

5) Choice D is correct

$Percent\ of\ change = \frac{new\ number - original\ number}{original\ number} = \frac{18.36 - 15.30}{15.30} = 20\%$

6) Choice B is correct

First find the multiples of 8 that fall between 10 and 60: $16, 24, 32, 40, 48, 56$. Since the greatest common factor of 32 and x is 8, x cannot be 32 (otherwise the GCF would be 32, not 8). There are 5 remaining values: $16, 24, 40, 48$ and 56. Number 16 is also not possible (otherwise the GCF would be 16, not 8). Then, there are 4 possible values for x.

7) Choice C is correct

The total number of candies in the box is $6 + 4 + 3 = 13$. The number of candies that are not orange is $6 + 3 = 9$. The probability of the first candy not being orange is $\frac{9}{13}$. Now, out of 12 candies, there are 8 candies left that are not orange. The probability of the second candy not being orange is $\frac{8}{12}$. Multiply these two probabilities to get the solution: $\frac{9}{13} \times \frac{8}{12} = \frac{72}{156} = \frac{24}{52} = \frac{6}{13}$

8) Choice B is correct

First, change the improper fractions into mixed numbers: $\frac{7}{2} = 3\frac{1}{2}$ and $\frac{30}{4} = 7\frac{1}{2}$

The integers between these two values are $4, 5, 6$ and 7. So, there are 4 integers between $\frac{7}{5}$ and $\frac{30}{4}$.

9) Choice C is correct

The increase in sales tax percentage is $8.5\% - 8.0\% = 0.5\%$

0.5% of \$250 is $(0.5\%)(250) = (0.005)(250) = 1.25\$$

10) Choice C is correct

When a point is reflected over x axes, the (y) coordinate of that point changes to $(-y)$ while its x coordinate remains the same. $C(7, 9) \rightarrow C'(7, -9)$

11) Choice B is correct

First, list the products:

$1 \times 4 = 4$

$1 \times 6 = 6$

$1 \times 5 = 5$

$1 \times 7 = 7$

$4 \times 6 = 24$

$4 \times 5 = 20$

$4 \times 7 = 28$

$6 \times 5 = 30$

$6 \times 7 = 42$

$5 \times 7 = 35$

Out of 10 results, 3 numbers are odd. The answer is: $\frac{3}{10}$

12) Choice C is correct

Since $5n$ is even, then $5n + 1$ must be odd. Thus $5n + 3$ and $5n + 5$ are also odd. So, there are a total of 3 numbers in this range that are odd.

13) Choice B is correct

The height of the building is 4 units.

$2{,}760\ centimeters \div 4 = 690\ centimeters$

14) Choice C is correct

First, use the given information to calculate the value of a: $\frac{a}{4} = b \rightarrow \frac{a}{4} = 2 \rightarrow a = 8$

Now, calculate $a^2 + 4b$ by substituting $a = 8$ and $b = 2$, $(8)^2 + 4(2) = 72$

15) Choice A is correct

Since 0.0099 is equal to 0.99%, the closest to that value is 0.1%.

16) Choice C is correct

Surface Area of a cylinder $= 2\pi r(r + h)$,The radius of the cylinder is 3 $(6 \div 2)$ inches and its height is 8 inches.

Therefore, Surface Area of a cylinder $= 2\pi(3)(3 + 8) = 66\pi\ in^2$

17) Choice C is correct

First, find the speed of the train in miles per hour: $280 \div 4 = 70$ miles per hour

The number of miles left to travel is: $1{,}500 - 280 = 1{,}220$ miles

To find the number of hours left, use the equation

$d = rt \rightarrow (distance) = (rate) \times (time) \rightarrow 1{,}220 = 70t$

$t = \frac{1{,}220}{70} = 17.4285714$ hours. That number rounded to the nearest whole hour is 17 hours.

18) Choice B is correct

Let x be the original price. Then:

$$\$1{,}912.50 = x - 0.15(x) \rightarrow 1{,}912.50 = 0.85x \rightarrow x = \frac{1{,}912.50}{0.85} \rightarrow x = 2{,}250$$

19) Choice C is correct

Let x be the number of inches representing 2.5 feet. Set up a proportion and solve for x: $\frac{x}{2.5} = \frac{0.15}{150} \rightarrow x = \frac{0.15 \times 2.5}{150} \rightarrow x = 0.0025\ in$

20) Choice D is correct

Set an equation: $\frac{3}{7}Z = 54$

Solve for Z: $\rightarrow Z = 54 \times \frac{7}{3} = 126$, then, calculate $\frac{2}{5}Z$: $\frac{2}{5} \times 126 = 50.4$

21) Choice A is correct

To answer this question, multiply 72 miles per hour to 8 $\rightarrow 72 \times 8 = 576$ miles

22) Choice D is correct

The spent amount is \$60, and the tip is 25%. Then: $tip = 0.25 \times 60 = \$15$

Final price = Selling price+tip $\rightarrow$ final price $= \$60 + \$15 = \$75$

23) Choice E is correct

To find the smallest factor of 95, list the factors: $1, 5, 19$, and 95.The smallest factor (other than 1) is 5. Of the choices listed $(28, 32, 39$, and $45)$, only 45 is a multiple of 5.

24) Choice A is correct

Adding exponents is done by calculating each exponent first and then adding and dividing:

$$\frac{4^2 + 3^2 + (-5)^2}{(9 + 10 - 11)^2} = \frac{16 + 9 + 25}{(8)^2} = \frac{50}{64} = \frac{25}{32}$$

25) Choice B is correct

Angle A and angle B are supplementary, so the sum of their angles is 180°.

Let a equal the measure of angle A, and let b equal the measure of angle B.

$a + b = 180$

The measure of angle A is 2 times the measure of angle B.

$a = 2b \rightarrow 2b + b = 180 \rightarrow 3b = 180 \rightarrow b = \frac{180}{3} = 60$

$a = 2b = 2(60) = 120$

Therefore, the measure of angle A is 120°.

26) Choice D is correct

First, convert their heights from feet and inches to inches, by multiplying the number of feet by 12 and adding the inches. Tomas:

6 feet +8.5 inches. 6(12 inches) +8.5 inches = 72 inches +8.5 inches = 80.5 inches

Alex: 5 feet +3 inches. 5(12 inches)+3 inches = 60 inches +3 inches = 63 inches

Then, subtract Alex's height from Tomas's height: $80.5 - 63 = 17.5$

27) Choice D is correct

So far, Kylie has written $10\% + 18\% = 28\%$ of the entire homework. That means she has $100\% - 28\% = 72\%$ left to write. $72\% = \frac{72}{100} = \frac{18}{25}$

28) Choice E is correct

Let x be the number of yellow pens. Write a proportion and solve: $\frac{yellow}{blue} = \frac{2}{3} = \frac{x}{9}$

Solve the equation: $18 = 3x \rightarrow x = 6$

29) Choice B is correct

To find the decimal equivalent to $-\frac{6}{9}$, divide 6 by 9. Then: $-\frac{6}{9} = -0.66666 \ldots = -0.\overline{6}$

30) Choice E is correct

The formula for the area of the circle is πr^2 , The area is 81π. Therefore:

$A = \pi r^2 \Rightarrow 81\pi = \pi r^2$, Divide both sides by π: $81 = r^2 \Rightarrow r = 9$, Diameter of a circle is $2 \times radius$. Then: $Diameter = 2 \times 9 = 18$

31) Choice E is correct

Five years ago, Amy was three times as old as Mike. Mike is 10 years now. Therefore, 5 years ago Mike was 5 years. Five years ago, Amy was: $A = 3 \times 5 = 15$, Now Amy is 20 years old: $15 + 5 = 20$

32) Choice B is correct

List the factors of 68: 1 and 68, 2 and 34, 4 and 17. There is one factor greater than 26 and less than 60.

33) Choice B is correct

Let the lengths of two sides of the parallelogram be $2x\ cm$ and $3x\ cm$ respectively. Then, its perimeter $= 2(2x + 3x) = 10x$

Therefore, $10x = 40 \rightarrow x = 4$

One side $= 2(4) = 8\ cm$ and other side is: $3(4) = 12\ cm$

34) Choice E is correct

Isolate x in the equation and solve. Then:

$\frac{3}{4}(x-2)=3\left(\frac{1}{6}x-\frac{3}{2}\right)$, expand $\frac{3}{4}$ and 3 to the parentheses $\rightarrow \frac{3}{4}x-\frac{3}{2}=\frac{1}{2}x-\frac{9}{2}$. Add $\frac{3}{2}$ to both sides: $\frac{3}{4}x-\frac{3}{2}+\frac{3}{2}=\frac{1}{2}x-\frac{9}{2}+\frac{3}{2}$. Simplify: $\frac{3}{4}x=\frac{1}{2}x-3$. Now, subtract $\frac{1}{2}x$ from both sides: $\frac{3}{4}x-\frac{1}{2}x=\frac{1}{2}x-3-\frac{1}{2}x$. Simplify: $\frac{1}{4}x=-3$. Multiply both sides by 4:

$(4)\frac{1}{4}x=-3(4)$, simplify $x=-12$

35) Choice C is correct

To answer this question, assign several positive and negative values to x and determine what the value of the expression will be:

x	-1	0	2	3	4
$2-x^2$	1	2	-2	-7	-14

So, the maximum value of the expression is 2.

36) Choice C is correct

The total number of handballs in the container is $6+5+8+10=29$. Since there are 6 green handballs, the probability of selecting a green handball is $\frac{6}{29}$.

37) Choice D is correct

If Emma answered 9 out of 45 questions incorrectly, then she answered 36 questions correctly. $\frac{36}{45}\times 100=80\%$

38) Choice C is correct

Let x be the number. Write the equation and solve for x.

30% of $x=12 \Rightarrow 0.30x=12 \Rightarrow x=12\div 0.30=40$

39) Choice A is correct

Let y be Anna's age: $5y+3=x \rightarrow 5y=x-3 \rightarrow y=\frac{x-3}{5}$

40) Choice B is correct

The area of a trapezoid can be determined using the formula: $A=\frac{1}{2}\times(a+b)\times h$

We know: $DC=4\ cm$, $AE=2\ cm$, and $AD=4\ cm \rightarrow$

$A=\frac{1}{2}\times(4\ cm+2\ cm)\times 4\ cm=12\ cm^2$

Effortless Math's GACE Math Online Center

... So Much More Online!

Effortless Math Online GACE Math Center offers a complete study program, including the following:

- ✓ Step-by-step instructions on how to prepare for the GACE Math test
- ✓ Numerous GACE Math worksheets to help you measure your math skills
- ✓ Complete list of GACE Math formulas
- ✓ Video lessons for Math topics
- ✓ Full-length GACE Math practice tests
- ✓ And much more...

No Registration Required.

Visit **EffortlessMath.com/GACE** to find your online GACE Math resources.

Our Most Popular Books!

Download at

Download at

Download at

Download at

Download at

Download at

Author's Final Note

I hope you enjoyed reading this book. You've made it through the book! Great job!

First of all, thank you for purchasing this study guide. I know you could have picked any number of books to help you prepare for your GACE Math test, but you picked this book and for that I am extremely grateful.

It took me years to write this study guide for the GACE Math because I wanted to prepare a comprehensive GACE Math study guide to help test takers make the most effective use of their valuable time while preparing for the test.

After teaching and tutoring math courses for over a decade, I've gathered my personal notes and lessons to develop this study guide. It is my greatest hope that the lessons in this book could help you prepare for your test successfully.

If you have any questions, please contact me at reza@effortlessmath.com and I will be glad to assist. Your feedback will help me to greatly improve the quality of my books in the future and make this book even better. Furthermore, I expect that I have made a few minor errors somewhere in this study guide. If you think this to be the case, please let me know so I can fix the issue as soon as possible.

If you enjoyed this book and found some benefit in reading this, I'd like to hear from you and hope that you could take a quick minute to post a review on the book's Amazon page.

I personally go over every single review, to make sure my books really are reaching out and helping students and test takers. Please help me help GACE Math test takers, by leaving a review!

I wish you all the best in your future success!

Reza Nazari

Math teacher and author

Made in the USA
Columbia, SC
10 November 2024